COMPUTER SCIENCE, TECHNOLOGY AND APPLICATIONS

ORACLE SQL FOR SECURE RELATIONAL DATABASES

COMPUTER SCIENCE, TECHNOLOGY AND APPLICATIONS

Additional books and e-books in this series can be found on Nova's website under the Series tab.

COMPUTER SCIENCE, TECHNOLOGY AND APPLICATIONS

ORACLE SQL FOR SECURE RELATIONAL DATABASES

RICHARD EARP
AND
SIKHA BAGUI

Copyright © 2021 by Nova Science Publishers, Inc.

All rights reserved. No part of this book may be reproduced, stored in a retrieval system or transmitted in any form or by any means: electronic, electrostatic, magnetic, tape, mechanical photocopying, recording or otherwise without the written permission of the Publisher.

We have partnered with Copyright Clearance Center to make it easy for you to obtain permissions to reuse content from this publication. Simply navigate to this publication's page on Nova's website and locate the "Get Permission" button below the title description. This button is linked directly to the title's permission page on copyright.com. Alternatively, you can visit copyright.com and search by title, ISBN, or ISSN.

For further questions about using the service on copyright.com, please contact:
Copyright Clearance Center
Phone: +1-(978) 750-8400   Fax: +1-(978) 750-4470   E-mail: info@copyright.com

NOTICE TO THE READER

The Publisher has taken reasonable care in the preparation of this book, but makes no expressed or implied warranty of any kind and assumes no responsibility for any errors or omissions. No liability is assumed for incidental or consequential damages in connection with or arising out of information contained in this book. The Publisher shall not be liable for any special, consequential, or exemplary damages resulting, in whole or in part, from the readers' use of, or reliance upon, this material. Any parts of this book based on government reports are so indicated and copyright is claimed for those parts to the extent applicable to compilations of such works.

Independent verification should be sought for any data, advice or recommendations contained in this book. In addition, no responsibility is assumed by the publisher for any injury and/or damage to persons or property arising from any methods, products, instructions, ideas or otherwise contained in this publication.

This publication is designed to provide accurate and authoritative information with regard to the subject matter covered herein. It is sold with the clear understanding that the Publisher is not engaged in rendering legal or any other professional services. If legal or any other expert assistance is required, the services of a competent person should be sought. FROM A DECLARATION OF PARTICIPANTS JOINTLY ADOPTED BY A COMMITTEE OF THE AMERICAN BAR ASSOCIATION AND A COMMITTEE OF PUBLISHERS.

Additional color graphics may be available in the e-book version of this book.

Library of Congress Cataloging-in-Publication Data

Names: Earp, Richard, 1940- author. | Bagui, Sikha, 1964- author.
Title: Oracle SQL for secure relational databases / Richard Earp (author), emeritus associate professor, Department of Computer Science, University of West Florida, Pensacola, FL, USA, Sikha Bagui (author), professor, Department of Computer Science, University of West Florida, Pensacola, FL, USA.
Description: New York : Nova Science Publishers, [2021] | Series: Computer science, technology and applications | Includes bibliographical references and index. |
Identifiers: LCCN 2021015004 (print) | LCCN 2021015005 (ebook) | ISBN 9781536194364 (paperback) | ISBN 9781536194807 (adobe pdf)
Subjects: LCSH: SQL (Computer program language) | Database security. | Relational databases. | Querying (Computer science)
Classification: LCC QA76.73.S67 E265 2021 (print) | LCC QA76.73.S67 (ebook) | DDC 005.75/6--dc23
LC record available at https://lccn.loc.gov/2021015004
LC ebook record available at https://lccn.loc.gov/2021015005

Published by Nova Science Publishers, Inc. † New York

Dedicated to my late wife, Brenda,
and
our children, Beryl, Rich, Gen and Mary Jo

And a special dedication to my wife, Anne,
who generously spent hours of her time editing this book
R.E.

Dedicated to my
father, Santosh Saha, mother, Ranu Saha,
husband, Subhash Bagui,
sons, Sumon Bagui and Sudip Bagui,
and nieces, Priyashi Saha and Piyali Saha
S.B.

Contents

Preface		ix
Chapter 1	Good Database Design	**1**
	1.1. Introduction	*1*
	References	*24*
Chapter 2	Database Integrity	**25**
	2.1. Introduction – What Is Integrity?	*25*
	References	*40*
Chapter 3	The Database	**41**
	3.1. Introduction	*41*
	References	*58*
Chapter 4	Privileges and ROLEs	**59**
	4.1. Introduction	*59*
Chapter 5	The Dictionary	**73**
	5.1. The Dictionary Paradigm	*73*
	5.2. Drilling Down into Information in the Dictionary	*74*

Chapter 6	Accessing Other Users' Tables with Scripts	**81**
	6.1. Backups	81
	6.2. Auditing	82
	6.3. Granting DML Privileges to Level 3 Users	83
	6.4. Using GRANT ALL	83
Chapter 7	Expanding the Database	**87**
	7.1. Getting Started with Linking Tables	88
	7.2. Integrity Constraints	93
	7.3. Some Beginning Queries	93
	7.4. Auditing Queries and Backups – Version 2	97
	References	103
Chapter 8	Triggers to Enforce Auditing	**105**
	8.1. What Is a Trigger?	105
	8.2. An Embellished Trigger for Hardware City	111
	8.3. Before Triggers vs. CONSTRAINTs	116
	8.4. Trigger Caveats	118
	References	119
Chapter 9	Multiple Users and Transactions	**121**
	9.1. Serialized Transactions	121
	9.2. COMMIT and ROLLBACK	125
	9.3. More on COMMIT and ROLLBACK	130
	9.4. SAVEPOINTs and ROLLBACK TO	131
	References	143
Chapter 10	Concurrency	**145**
	10.1. Consistency, Concurrency, and ACID Properties	145
	10.2. Transactions Revisited	147
	10.3. LOCKing	153
	10.4. Deadlocks and Starvation	156
	References	167
About the Authors		**169**
Index		**171**

PREFACE

A typical Oracle database has multiple users working simultaneously. Data is shared amongst the users, and this leads to security concerns. This book comes in from the angle of developing and maintaining a secure Oracle database with multiple users. In this book, we assume that you are acquainted with basic Oracle SQL and fundamentals of relational database.

While a fundamental tenet of database is the sharing of data, the data to be shared must be "good" data. To ensure goodness, we review database design in Chapter 1 to solidify the rationale and process required to reach the third normal form. Chapter 2 reviews constraints to ensure integrity but the focus of the material is still on the single user. In Chapter 3, we begin working as multi-users; we embark on sharing one good database.

This book relies heavily on examples and scenarios and assumes you will do the exercises in each chapter. We will often suggest doing something within Oracle that is not "standard" but rather what we consider "good practice." Our background includes a heavy dose of software engineering, and our experience influences the way we approach database topics. In dealing with secure multi-user environments, it is imperative to follow practices making the software we write obvious and clear. For example, we try to include meaningful comments in most queries with several lines; we also advocate fully parenthesized expressions for arithmetic and logic clauses.

The majority of money and time spent on software is in maintaining it. Very often, multiple people are looking after databases and software written by someone else. Someone other than the original author of a particular query or a table creation has to change or adapt the code. The maintainer should never have to guess what the original author had in mind when the commands were written. There are rules governing operator precedence, but not everyone knows those rules; and it is, in our opinion, clearer to write queries without relying on precedence rules.

The same rationale applies to commenting. It is most helpful to maintainers if they know who wrote the original query, when it was written, what each non-obvious attribute means, and what the query should accomplish.

To deal with multiple users in a secure Oracle database environment, we must define some terms to use in discussing individual responsibilities within the organization and within the database. After reviewing some basic ideas in database, we begin the secure multi-user part with a description of privileges and tablespaces -- what a person with an Oracle account can and cannot do with the database and what data can and cannot be seen. We will create a tablespace and the Data Base Administrator manages it. The secure multi-user part of book begins in earnest in Chapter 3.

The running example is a small business where there is a business owner and "computer people," as well as other employees who work in various departments such as sales, purchasing, human resources, inventory management, and accounting. We are going to follow these individuals through the development of a secure database. How can sharing of data be performed in an orderly fashion? What is a good secure database?

Chapter 1

GOOD DATABASE DESIGN

1.1. INTRODUCTION

One of the most important ideas for using a centralized database is to devise a way to share good data. There are two points to be made about sharing. First, sharing data requires rules for how data will be shared and by whom. But, before we worry about sharing data, the other point about sharable data is it must be "good."

What makes good data? And, what do we mean by a good database? The answer to these two questions resides in database design and the enforcement of integrity rules. We will review the points of database design first, and then we will address database integrity in Chapter 2. We begin the discussion of sharing data in Chapter 3.

1.1.1. Database Design

A database is about all people in an enterprise using one and only one version of data. In the early days of computers, a company may have had numerous programmers who kept data for their department or division. For the moment, suppose there were three programmers in a company -- one in the Billing department, one in Sales, and one in Inventory. Suppose each of these individual programmers kept a file about the products and customers who bought the products. Billing needed to know the price paid for the products and what to charge customers who bought the product. Sales needed the same kind of information and also the quantity-on-hand and customer information for future sales. Inventory needed not only quantity-on-hand but also order points for inventory items and supplier information.

Many years ago, the concept of one file (or better yet, one set of files), one programmer in one department was called the "my file" concept. Imagine a large company with numerous programmers each keeping their own version of product data for their department. You might well expect when taken as a whole, the idea of "product data" depended on which programmer's files were current and accurate.

The problem with the "my file" idea is this plethora of files kept by each group would probably have stored data redundantly. If one wanted to know how many things were in inventory, it depended on who was asked. If one wanted to know a customer's address, it again depended on who supplied the information and how current it was. Although it might be expected two independent programmers would keep their data current, more than likely they updated their data in different ways at different times. So, early in the days of computers and files, the question was asked, "Why not share **one** file about products or customers or inventory across the whole company?" Then, all departments and their personnel would be accessing the same information.

Sharing data is a good idea; but, before we share data, we need to discuss the "goodness" of the data to be shared. Our first consideration in regarding goodness of data is to define relational normal forms and use them for our database. While special cases may exist where one picture of data does not seem to fit every situation, generally a relational database in the third normal form is considered a good database. We shall now see how to get to the third normal form.

1.1.2. The First Normal Form (1NF)

When we look into the creation of a database, we must first consider the design of it. If a person wants to build a house, the first thing to do is to draw up a plan, a blueprint. Building a database correctly must be done is a similar way. What data goes with which table? Are there right and/or wrong ways to design tables? *Relational database* consists of creating two-dimensional tables -- tables with rows and columns. The two-dimensional arrangement of data mirrors the concept of matrices in mathematics. A two-dimensional mathematical matrix has rows and columns like this:

Matrix

	Column1	Column2	Column3
Row1	33	11	66
Row2	22	66	33
Row3	99	33	44

Relational database was introduced by E. Codd (see References at the end of the chapter). Codd envisioned database in mathematical terms and created the notion of normal forms. As you will see, data in relational databases look very similar to mathematical matrices. For example, a sample of a table of data about classes in a school and the rooms in which classes are offered might look like this:

Rooms

	Math	Chemistry	Physics
8 AM	10	12	14
9 AM	12	16	10
10 AM	14	10	12

A Chemistry class is offered at 9AM in room 16.

Codd's concept of a data matrix led to his creation of a set of rules for a good relational database. Our first venture into relational database involves realizing all data can be arranged in two dimensional tables. Further, the data in the tables must be arranged correctly to be efficiently searched.

Three normal forms are defined in describing data in two dimensional tables. These arrangements of data are referred to as the first, second, and third normal forms. A correct relational database is at least in the third normal form (3NF). Having a database in the 3NF implies data is also in the first (1NF) and the second normal forms (2NF). In the next few sections, we explain the principles of normal forms. As you peruse the ideas presented in normalizing a database, the traditional approach is to define the normal form (NF) as what the NF does not contain. So, bear with the descriptions as they tend to define things such as, "If table R does not contain X, then R is in normal form Y."

The first normal form (1NF) demands all data in tables be "atomic." Codd originally used an expression describing values of attributes as from a "simple domain." Atomicity or a "simple domain" implies the data items cannot be broken down -- hence, the characterization of the data as "atomic." There are two ways related data can be non-atomic -- repeating groups and composite attributes. A table in 1NF has only atomic attributes; it does not have repeating groups or composite attributes.

1.1.2.1. Repeating Groups

Suppose a table contained data about employees in a company and the departments in which they are qualified to work. Consider this arrangement of a row of data:

<101, Alice Smith, Sales, Finance, Marketing>

This data implies the **Employee** table has this form:

Employee (EmpNo, EName, Dept1, Dept2, Dept3, ...)

Or

Employee (EmpNo, EName, {Dept})

EmpNo is the title of the column of data for employee numbers.

EName will title the column of employee names.

Dept is represented as {Dept} to illustrate it contains multiple occurrences of Departments in which the employee may be qualified to work.

What is wrong with this design?
What is wrong is finding a department and matching that department to a given employee is at best difficult. {Dept} or Dept$n$, $n = 1, 2, 3,...$ is in a "repeating group." A few rows in the **Employee** table as defined above might look like this:

Employee

EmpNo	EName	Dept1	Dept2	Dept3
101	Alice Smith	Marketing	Finance	Sales
102	Bob Baker	Finance	H.R.	
103	Chuck Charles	Marketing	Sales	

Now, suppose you want to find all the employees who are qualified to work in Sales. How do you find this information in the arrangement of data above? You have no choice but to look at every row in the database and see if an employee has the qualification you seek. In this design qualifying departments are listed for each employee and no order of the departments within {*Dept*} is implied. Even if the departments were in alphabetical order for each employee, difficulties still arise.

The *Sales* department occurs in different places in the repeating group in different rows. Even if the repeating group were in alphabetical order, we see *Sales* occurs in the third position of the first row and in the second position of the third row. Repeating groups are not readily searchable as the design stands.

If an employee were to qualify for a new department, adding that data to the table in this alphabetical arrangement of departments may be a problem. Consider adding the qualification of *H.R.* (Human Resources) to employee 103. Adding this data requires a reorganization of the {Dept} data in 103's row to put *H.R.* first in 103's alphabetically organized list of qualifications and moving the other data to make room for the new entry.

Data such as presented above with the repeating group is said to be "not in the first normal form." The first normal form (1NF) disallows repeating groups because the data in the repeating group is not readily searchable and may cause update difficulties. To rid the database of repeating groups, it is put in a new table along with a reference to the original table. The table **Employee** should be rearranged into two tables like this:

Employee (EmpNo, EName)

Qualifications (Dept, EmpNo)

Such a rearrangement is called "decomposition." The idea of decomposing a table to move to a better arrangement of data is very common in relational database. The data from above in two tables would look like this:

Employee

EmpNo	EName
101	Alice Smith
102	Bob Baker
103	Chuck Charles

Qualifications

Dept	EmpNo
Marketing	101
Finance	101
Sales	101
Finance	102
H.R.	102
Marketing	103
Sales	103

Why is the latter design better than the original arrangement? First of all, the data in the **Qualifications** table is now accessible by *Dept*. If you were looking for a qualified employee for *Marketing*, you need only search the rows of the **Qualifications** table which would point you to the *EmpNo*s for those qualified for *Marketing*. You might say, "Before we had to look at all rows of the original **Employee** table to find *Marketing*, and now we have to look at even more rows in the **Qualifications** table to find it."

The difference is the *Dept* attribute in the **Qualifications** table can be arranged to make finding specific job titles easier to find. One arrangement would be to alphabetize the job titles. If you were looking for *Marketing*, the alphabetical list of job titles in the **Qualifications** table would make finding *Marketing* easier. Another facility for finding data in **Qualifications** would by indexing the *Dept* column.

Codd went a step further in defining relational database to be sets of atomic data arranged in matrix-like tables. The keyword in the previous sentence is "sets." Mathematical sets have no implied order. The above **Employee** table could be depicted as:

Employee
EmpNo	*EName*
103	Chuck Charles
101	Alice Smith
102	Bob Baker

The two representations of the **Employee** table are equivalent *sets* of data. In sets, things are in the set or they are not. Where things are in the set is not defined. In relational database, *tables are sets of rows*. When a table is defined, the arrangement of columns is fixed – every row will have an *EmpNo* first, then an *Ename*. When the table is populated, no sense of which row comes first or second or … is ever assumed.

Reconsidering the problem of adding a qualification of *H.R.* to EmpNo *103* is greatly simplified in the decomposed version of the table **Qualifications**. The row:

<H.R.,   103>

is added to the table of **Qualifications**. It does not matter where the row is placed in the set of **Qualifications** because the order of the rows is not defined. Adding the new row is straightforward as no further re-arrangement is required as it could have been with the repeating group.

1.1.2.2. Composite Data Items

Suppose our **Employee** table represented a large company located in several cities. Each employee's principal work site is located in some city and state. We add a *Location* (City and State) to the **Employee** table, and the new table design is:

Employee (EmpNo, EName, Location)

Here is a sample of the **Employee** table with some data:

EmpNo	EName	Location
101	Alice Smith	Mobile, AL
103	Chuck Charles	Tampa, FL
102	Bob Baker	Pensacola, FL

Based on the sample data, we notice *Location* contains two parts -- a city and a state. The city-state combination is called a "composite" attribute because it is composed of two parts. If there were never a question about finding employees by state, this arrangement of data could possibly be acceptable. However, the city and state should be entered as separate attributes so no assumption about how data should be accessed is broached. A more proper representation of our **Employee** table would be:

Employee (EmpNo, EName, Location (City, State))

Or, better yet:

Employee (<u>EmpNo</u>, EName, Location.City, Location.State)

The sense of *Location* is "where the person works" rather than a home address. If the sense of *Location* were understood, the qualifier in the table description could be dropped.

There could be situations where someone must find all the employees working in a state. For example, a person in the payroll department might need to calculate a state income tax rate for each employee. If the data were arranged as we have it above, finding employees by state would again require a row-by-row search of our **Employee** table. As you look at examples, think beyond the example to a table with thousands of employees.

Here, *Location* is called a *composite* attribute -- it consists of parts which compose the whole attribute *Location* -- a *City* and a *State*. Tables with composite attributes are considered non-atomic because the attribute can be split into parts. With the composite broken down, and the meaning of *Location* clear, a better representation of the **Employee** table would be:

Employee (<u>EmpNo</u>, EName, City, State)

Our table with sample atomic data would now look like this:

Employee

EmpNo	EName	City	State
103	Chuck Charles	Tampa	FL
101	Alice Smith	Mobile	AL
102	Bob Baker	Pensacola	FL
104	Donna Davis	Pensacola	FL

Although the data above seems to be atomic, the composite attribute has left us with a column that is less "searchable." The person in payroll would prefer the *State* be split off to focus on finding employees in states so what do we do? Decompose the table to arrange the data like this:

Employee

EmpNo	EName	City
103	Chuck Charles	Tampa
101	Alice Smith	Mobile
102	Bob Baker	Pensacola
104	Donna Davis	Pensacola

State

State	EmpNo
FL	103
AL	101
FL	102
FL	104

The 1NF is defined as: All data in a table must be *atomic*. Atomicity means we rid the table of *repeating groups* and *composite structures.* How do we reorganize non-1NF tables? We decompose the original table into tables not containing the offending (non-1NF) parts.

1.1.3. Keys

What is a key in the sense of a database? A key is an attribute in a table that identifies another attribute. In our example, knowing an employee number, *EmpNo*, can be used to identify information about an employee. *EmpNo* is a key.

Looking at the latest **Employee** table, we have three attributes:

Employee (EmpNo, EName, City)

Knowing an *EmpNo* will lead to finding an *Ename* and a *City*. Would knowing an employee's name identify an *EmpNo*? Maybe, maybe not. Some people have the exact same name and hence *Ename* is not necessarily unique enough to identify an *EmpNo*. *EmpNo* is defined to be *unique* for each employee. The argument for and against *Ename* being a key is a semantic one. The point that *Ename* may not be unique and hence is not a good candidate for a key is based on the possibility of people having the same name.

City is likewise not a good candidate for identifying either the *EmpNo* or the *Ename*. Just looking at the sample data, finding an employee by knowing where the employee works (*City*) is not going to identify a specific employee. Disqualifying an attribute as a key candidate by inspecting data in a table is called a "counter-example argument." Find two rows where knowing *City* implies knowing the *EmpNo* rules out *City* as a candidate key.

Every table in relational database has a primary key. The primary key of a table must be unique so as to identify all the information in a row of the table. In the **Employee** table, *EmpNo* is defined to be a unique row identifier. If an *EmpNo* is supplied, one row in the table is associated with that attribute; *EmpNo* is the primary key of the **Employee** table. Because *EmpNo* is the primary key, we underline it in the description of the table like this:

Employee (<u>EmpNo</u>, EName, City)

Now consider the **Qualifications** table:

Qualifications (Dept, EmpNo)

What is the key of this table? It can't be *Dept* because there are several departments associated with various employees. *Dept* is not a unique identifier. How about *EmpNo*? In the **Qualifications** table, *EmpNo* is not unique either as employees with multiple qualifications occupy more than one row. Since both attributes are disqualified candidate keys, the only possibility for a key of **Qualifications** is the concatenation of *Dept* and *EmpNo*.

Every table in relational database has a unique primary key. Why is this true? In a relational database, tables are sets of rows -- just like a mathematical set. In mathematical sets, things are either in the set or they are not; but, there is no sense of ordering and no duplicate entries. We said,

"Every table in relational database has a primary key." This is true because if you take the combination of all the attributes in a table, the row where those attributes appear is part of a set and hence must be unique.

The actual primary key of a table may consist of fewer attributes than all of them. Consider the example of **Employee**:

Employee

EmpNo	EName	City
103	Chuck Charles	Tampa
101	Alice Smith	Mobile
102	Bob Baker	Pensacola
104	Donna Davis	Pensacola

Does the combination <101, Alice Smith, Mobile> uniquely define a row? Of course it does because there are no duplicate rows. This is a *set* of rows. Every row is unique; therefore, if you consider the concatenation of all attributes values as a key, it will always work. On the other hand, considering all attributes concatenated together as a primary key could be a bit of an overkill. We will look again at the idea of minimal keys after we introduce functional dependencies in the next section.

1.1.4. Functional Dependency

A functional dependency (FD) is way of describing "what identifies what." To further discuss more normal forms, we will find them easier to define using functional dependencies. An example of a FD would be if finding a name in the **Employee** database depends on specifying an employee number, then an employee number identifies a name. Since knowing an *EmpNo* identifies an employee's name, we say *Ename* is functionally dependent on *EmpNo*. The FD is written like this:

EmpNo -> EName

Good Database Design 13

This expression is read, *EName* is functionally dependent on *EmpNo*. You could say finding *EName* in the database depends on knowing an *EmpNo* or *EmpNo* identifies *Ename*.

1.1.4.1. Finding a PRIMARY KEY Using Functional Dependency

Let's return to the idea of the key once again. We stated every relational database has at least one key -- all of the attributes concatenated together. From a functional dependency (FD) standpoint, if we look once more at the table,

Employee (<u>EmpNo</u>, EName, City, State)

we could identify *EName* with this FD:

EmpNo, EName, City, State -> EName.

One way to find a *minimal* key from the obvious FD of all attributes concatenated together is to begin testing each attribute and asking if the FD holds or not. With **Employee**, we ask these questions:

Does *EmpNo* -> *EName*?

It does, because *EmpNo* is defined to be unique for each employee.

We then look at the FD, *EmpNo* -> *City*. The sense of the FD has to be considered and what this FD means is a given employee works in a City somewhere. Does the proposed FD, *EmpNo* -> *City*?

It does, because *EmpNo* is unique and the employee is assigned to work in one city. If an *EmpNo* is given, one finds a location assigned to the employee in this table.

So, since *EmpNo* identifies all other attributes in the table and we defined it as unique, we elevate the status of *EmpNo* and call it a "candidate key."

Each of the other attributes is considered semantically as to its fitness as a candidate key or by looking at sample data in the populated table.

Does *EName -> EmpNo*? No, because there could be employees with the same name. Semantics tells us *EName* is unsuitable as a candidate key.

Does *State -> EmpNo*? No, a state cannot identify an *EmpNo* because a *State* (in the *Location* composite) is likely assigned to more than one employee. Semantics disqualifies *State* as a candidate key as does finding a counter example. If data were supplied for the table, as it is above, you have two employees assigned to work in Florida; hence, this alone disqualifies *State* as a candidate key.

As each scenario is examined, we find we have only one candidate key; hence, *EmpNo* is the primary key. A <u>primary key</u> is said to be a *chosen* candidate key and is underlined in a relation.

To illustrate the 2NF, we need to add some more data to our employee example.

We would like to add a salary. We assume the salary for a particular job qualification depends on the qualification itself as well as the employee working in that department. Where in the above database do you put salary?

1.1.4.2. In Which Table Do We Put Salary?

Before we answer this question definitely, we'd like to give some background thoughts. Where would we put the salary in the two tables we have defined so far? We cannot put the salary in the **Employee** table because it takes *both* the *EmpNo* and the *Dept* to indicate a salary. The functional dependency designation would look like this:

EmpNo, Dept -> Salary

Neither the *EmpNo* nor the *Dept* taken alone will identify a *Salary*. The description of the **Qualifications** table with *Salary* included looks like this:

Qualifications (<u>*Dept, EmpNo*</u>, Salary)

Here is some sample data:

Qualifications

Dept	EmpNo	Salary
Marketing	101	47
Finance	101	25
Sales	101	51
Finance	102	31
H.R.	102	39
Marketing	103	42
Sales	103	41

So what is the 2NF? The 2NF is defined as follows: All data in the table is dependent on the whole key of the table. The *Salary* attribute is defined by (dependent on) the concatenated key of *EmpNo* and *Dept*. Therefore, the **Qualifications** table is in the 2NF.

1.1.4.3. A Closer Look at Functional Dependency and Normal Forms

Functional dependency is often discussed in a semi-mathematical form using the letters R, S, … for tables, and A, B, C, … as the names of attributes.

Table R with primary key A and non-key attributes B and C is depicted as:

R(<u>A</u>BC) where A -> BC

The 2NF would be depicted as follows:
Table S exists with attributes A, B, and C. The key of S is AB, and C is functionally dependent on AB:

S(<u>AB</u>C) where AB -> C

To fulfill the 2NF, if an attribute D is included in S, then D must have AB -> D. If a FD, A -> D or B -> D, exists, then the table S is not in 2NF.

1.1.4.3.1. More FD Manipulations

In R(AB), given A -> B, does B -> A? No. Here is a counter example:

EmpNo -> EName because *EmpNo* is defined to be unique and an *EmpNo* is assigned to one and only one employee. Since names are not normally unique, the semantics of the situation suggests it would be incorrect to assume *EName -> EmpNo*.

In R(ABC), given AB -> C, does A -> C or B -> C? No.

To accept the conclusion A –> C, we would be saying every occurrence of C is identifiable by A. The **Qualifications** example may be used to illustrate the point. In **Qualifications**, *EmpNo* and *Dept* are both needed to identify a *Salary*. We know by semantics or examination of the data that A -> C or *EmpNo -> Salary* is not true. The first two rows of the table confirm this.

Qualifications

Dept	EmpNo	Salary
Marketing	101	47
Finance	101	25
Sales	101	51
Finance	102	31
H.R.	102	39
Marketing	103	42
Sales	103	41

1.1.4.3.2. Add an Address

Now suppose we add an employee's address to the employee database. Which of these tables would be correct and why?

Employee (<u>EmpNo</u>, EName, City, State, Employee_address)

R(<u>A</u>BCDE), A -> BCDE

or

Qualifications (<u>Dept, EmpNo</u>, Salary, Employee_address)

S(<u>AB</u>CD), AB -> C, A -> D

It is reasonable to assume an employee lives at some address. Hence, semantics tells us *EmpNo -> Employee_address*. In the **Employee** table, we can find an employee by the *EmpNo*, and it is logical to include *Employee_address* as an attribute in the table. The functional dependency tells us to put the address in **Employee** with the *EmpNo*.

Had we chose to put *Employee_address* with the **Qualifications** table, there is a difficulty. *EmpNo* is not a unique identifier in this table. The symbolic representations bear this out as they tell us AB -> C and A -> D which means if D is in S, then S is not in 2NF.

The 2NF requires all data in a table depend on the whole key of the table. If the address were put in the **Qualifications** table, we'd have a *partial dependency* -- the address would depend on only part of the key.

What is wrong with tables not being in the 2NF? The answer is *redundancy* and *anomalies*. An anomaly is when an action is taken but something undesirable is a side effect. Consider some sample data in the non-2NF table:

Qualifications (Dept, EmpNo, Salary, Employee_address)

Qualifications_X2NF

Dept	EmpNo	Salary	Employee_address
Marketing	101	47	23 Palafox St.
Finance	101	25	23 Palafox St.
Sales	101	51	23 Palafox St.
Finance	102	31	86 Vine Ave.
H.R.	102	39	86 Vine Ave.
Marketing	103	42	5 Park Place
Sales	103	41	5 Park Place

The non 2NF problem is obviated by the redundant entering of the address for employees appearing multiple times in the table.

An insert anomaly would be if someone wanted to add a new employee number and address. The table **Qualifications_X2NF** would require that an employee be assigned to a department or the data could not be entered. Why not? Because no primary key or part of a primary key can be *null* so a row like:

<null, 104,32,21957 Delta Rd.>

cannot exist in our database.

With the data presented above, let an employee's address be changed. Employee 101 moves from 23 Palafox St. to 65 State Blvd. Is the change in just one place? No, due to the redundant entries of address in this table, multiple updates are required. This redundant updating is an "update anomaly."

If Employee 102 is terminated, then the entries in **Qualifications_X2NF** dealing with 102 are deleted. The problem with this deletion is the employee's address is also deleted. Losing the information about this employee's address is an example of a delete anomaly. Had the employee's address been in the **Employee** table, it may have been possible to preserve the information about the employee's address by some means.

If we look at putting the *Employee_address* in the **Employee** table and adding the *Date_started* and *Date_terminated* attributes, the anomalies go away. If the **Employee** table had this form:

Employee (EmpNo, EName, City, State, Employee_address, Date_started, Date_terminated),

and if the original data for employee 112 was:

<112,   Tom Tomato, Grass Valley, CA, 45 Palafox St., 1 June 2015, null>

Here, all the attributes are identified by the employee number. Changing the address involves changing only one row in the **Employee** table. Terminating Tom involves changing the *null* value for *Date_terminated* to a real date. Adding Tom to the database without assigning a qualifying department or salary is no problem. The anomalies and redundancy are removed when tables are in 2NF.

1.1.4.4. Concatenated Keys and Intersection Data

When we see a concatenated key in a database, it is often a sign that an M:N relationship is in play. For example, in a school, one may have a **Student** table and a **Course** table. The relationship of students to courses is many to many (M:N) as many students take many courses and many courses are taken by many students. The intersection table relating these two entities might look like this:

Enrollment (Student_id, Course_id, Instructor, Building, Room, Grade)

The primary key of **Enrollment** is the concatenation of *Student_id* and *Course_id* which are presumably the keys of the **Student** and **Course** tables. The other attributes in this intersection table are called "intersection data" because their occurrence depends on the intersection of *Student_id* and *Course_id* in **Enrollment**.

The thing to examine when a concatenated key exists is whether or not all the attributes (the intersection data) truly depend on the whole key. If any intersection attribute depends on only part of the key, then the table is not in 2NF.

When we are satisfied that we have reached 2NF, we can go further with the normalization process in looking for violations of the third normal form.

1.1.5. The Third Normal Form, 3NF

For a table to be in 3NF, it is required that all attributes in the table depend principally on the primary key of the table. The current **Employee** table has this structure:

Employee (<u>EmpNo</u>, EName, City, State, Employee_address, Date_started, Date_terminated).

The functional dependency situation is:

EmpNo -> EName, City, State, Employee_address, Date_started, Date_terminated

The *EmpNo* is the primary key of the table and identifies all the attributes therein. Add two more attributes to the **Employee** table. Each employee is assigned one and only one project. The project is identified by a *Project_id* and is managed by a supervisor. We enhance our **Employee** table by adding the *Project_id* and a supervisor for the project.

Employee (<u>EmpNo</u>, EName, City, State, Employee_address, Date_started, Date_terminated, Project_id, Project_supervisor)

What about the functional dependencies? The original ones still hold:

EmpNo -> EName, City, State, Employee_address, Date_started, Date_terminated.

The added attributes may be identified by the *EmpNo* as well. Since there is only one project for each employee, the functional dependency:

EmpNo -> Project_id

is valid.

What about the *Project_supervisor*? It is true since the employee has only one project assigned and the supervisor for the project is also identifiable by the *EmpNo*. The problem here is the supervisor of the project is better identified by the *Project_id* than by the *EmpNo*. The better description of functional dependencies would be:

EmpNo -> Project_id

and

Project_id -> Project_supervisor

These two functional dependencies are called a transitive dependency. Symbolically, we have:

R(ABC...) where A -> B and B -> C.

The problem is that A is the key of R, and B depends on A; but C does not depend on A as much as it depends on B. The 3NF requires all attributes in a table depend on the key of the table. The **Employee** table in the above form containing *Project_supervisor* is said to be not in the 3NF.

The remedy for "not in the 3NF" is decomposition. If

R(ABC) where A -> B and B -> C,

decomposition is

R1(AB) and R2(BC).

Here, B depends on A in R1, and C depends on B in R2. R1 and R2 are now in 3NF. The **Employee** table is decomposed into:

Employee (EmpNo, EName, City, State, Employee_address, Date_started, Date_terminated, Project_id)

and

Project (Project_id, Project_supervisor)

There have been attempts to improve the 3NF in relational database where there are special cases of functional dependency. In practice, the 3NF seems to be sufficient in almost all cases of database design. We will leave the "beyond 3NF" as an academic exercise you may choose to pursue.

1.1.6. The Equi-join

Since we have decomposed the tables in the database, how are they put back together? The reconstruction is done with queries to recombine the decomposed tables. For example, answering a request like, "Find the employees supervised by Sam Smith," would call for an equi-join of **Employee** and **Project** like this:

```
SELECT    e.EmpNo, e.EName
FROM      Employee e, Project p
WHERE     p.Project_supervisor = 'Sam Smith';
```

Exercises for Chapter 1

Ex 1.1. Normalize the following to the 3NF:
   a. R(ABCD), AC -> D, A -> B
   b. R(AB{C}D), A-> BD, {C} is a repeating attribute
   c. R(ABC), ABC -> ABC
   d. R(ABCD), A -> BC, B -> D
   e. R(ABCDE), A-> BCD, C-> E

Ex 1.2. Indicate whether the following are true or false. Explain your rationale or provide an example of why or why not the FD holds. (These are functional dependencies, not normal forms.)
   a. If R(AB) and A -> B, then B -> A
   b. if R(ABC), A -> BC, then AB -> C
   c. If R(ABC), A -> B and B -> C, then A -> C
   d. if R(ABC), A -> BC, then A -> B and A -> CEx 1.3

Every table has a primary key -- True or False? Why?

Ex 1.3.
A database design problem:
Design a database such that all tables are in 3NF.

A person collects baseball cards. Each card has a wholesale and retail value. Each card is relevant to one and only one player, but the player may be on different teams at varying times. A given baseball card for a player is for the time when that player is part of some team. For example, John Jones was with the Pirates in 2016, but in 2020 and 2018 he played for the Tigers. Jones would have a card for each year for each team.

The player has a batting average and a count of home runs by the season for a team. Each player has a birth date, handedness (right or left), hitting preference (right, left, or both), home address, city, state, and zip. You also need a listing of teams with appropriate information (City, State, Stadium name, Mascot, Owner, General Manager).

It is suggested that you have your design approved by your instructor. If no instructor is available, explain your design to another person in a structured manner to see if your database is understood. Relationship should be described as one to many, many to many, or many to one (the relationship cardinality). And the description of a relationship should include the words may or must. (Is the relationship mandatory or not?). As an example, you might say, "A player *may* be related to *one or more* teams. Every team *must* have *one or more* players. A team *must* have *one or more* owners."

Describe how you envision the relationship of team to player. The relationship is likely many to many (M:N) so explain how your tables reflect the relationship and what intersection data is involved. Explain why your database is in 1NF, 2NF, and 3NF.

REFERENCES

Codd, E. F., (1970a). *Notes on a Data Sublanguage,* IBM internal memo (January 19, 1970).

Codd, E. F., (1970b). "A Relational Model of Data for Large Shared Data Banks," *CACM 13*, No. 6 (June, 1970).

Codd, E. F., (1970c). *The Second and Third Normal Forms for the Relational Model,* IBM technical memo (October 6, 1970).

Codd, E. F. (1971). *A Data Base Sublanguage Founded on the Relational Calculus,* IBM Research Report RJ893 (July 26th, 1971).

Chapter 2

DATABASE INTEGRITY

2.1. INTRODUCTION – WHAT IS INTEGRITY?

Our presentation of database integrity in this Chapter is still at the one-user level. We will begin looking at multi-users in Chapter 3. We are working toward sharing good data in a database. The point of sharing data is the central idea in database. Sharing is only useful if the database to be shared is designed correctly, as presented in Chapter 1 and if the shared data is trustworthy. In database we speak of this trustworthiness as *integrity*.

Database integrity means the information within the database can be trusted. If someone asked (queried) the database for a customer's address, we expect one result, one address. If the result of the query "What is Richard Walsh's address?" gave us multiple results, either one of two things is true:

(1) There are two people with the same name in the database; in which case, the database design needs to be refined to include a unique identifier for each customer.

or

(2) The database lacks integrity because you would be getting two answers when you expect one. Somewhere in the database this person has multiple addresses stored redundantly.

In either case, we would say the database lacked integrity. How do we ensure integrity in the database? How do we keep our data secure?

We begin by introducing constraints. Constraints are part of table definition. To ensure database integrity, we must have ways of insuring the stored data is entered correctly. All data should make sense; it should have limits on its acceptable domain of values. Further, data items are either unique or not. If a data item such as a unique customer number is to be truly unique, the design must insure this. If a data item such as an address can appear multiple times, it must be findable by some unique identifier.

In addition to the issue of integrity, the tables in the database must be constructed to ensure security. Data must meet predefined criteria. If data can be changed to disrupt integrity, steps must be taken to insure such changes are disallowed by restricting value changes as well as restricting users from tampering with data they should not change.

We begin the journey of creating a fault-free set of tables with our CREATE TABLE commands which reflect the correct design of the database. In this chapter we will be single users in default mode. In moving in the direction towards a multi-user environment, some of the defaults for single users will be modified as we enter Chapter 3. Here the security focus is on disallowing bogus values. Later, we will handle the security of the database by restricting the privileges of users.

2.1.1. The CREATE TABLE Command

When you first learned SQL, you learned to create tables from your account. Your first account was probably set up by an instructor and most likely involved simply signing on without consideration of space or privileges. Given the default environment, the simplest form of the CREATE TABLE involves naming the table, the attributes, and their datatypes. As an example for this chapter, we will create tables storing data about children in a kindergarten. While we will present the SQL to create the kindergarten database tables, first realize that some database design should have been completed. We begin with a CREATE TABLE looking like this:

```
CREATE TABLE Child(
    ID      CHAR(9),        -- Social Security Number
    Name    VARCHAR2(20),   -- Up to 20 characters
    Height  INT             -- Height in inches
);
```

Without thinking too hard about it, we have already made decisions affecting integrity. We have so far limited the number of attributes to only three. We have decided all children will be identified by an *ID* attribute which we defined as a character string of 9 alphanumeric characters. For the example, we could use social security numbers (which are now assigned at birth to citizens of the United States). We have limited the name to 20 characters and have decided to store the *Height* as an integer representing how tall a child is in inches.

In another country, the size and datatypes may be different, but we will continue to use this example.

2.1.1.1. Defining the PRIMARY KEY CONSTRAINT

Missing in this CREATE TABLE example are integrity rules preventing incorrect data from populating the table. Our first integrity rule is to define the primary key of the **Child** table. Primary keys are always unique and identify all the attributes in any row. Primary keys may be defined as the table is created or may be added after the fact using an ALTER TABLE command. The table creator should have designed the database to the point where using ALTER TABLE would be less likely for the PRIMARY KEY construction; nonetheless, we will illustrate all three ways to define the primary key.

The first way to define a primary key is very common for single attribute primary keys. The technique is to include the primary key constraint on the same line as the attribute. The SQL would look like this:

```
CREATE TABLE Child(
    /* CREATE TABLE PRIMARY KEY Version 1 */
    ID      CHAR(9)      PRIMARY KEY,
            -- Social Security Number
    Name    VARCHAR2(20),
            -- Char string up to 20 characters
    Height INT
            -- Height in inches (Integer)
);
```

The responsibility and management of comments fall to the table creator. We encourage meaningful comments.

A preferable way to add the PRIMARY KEY constraint is to add it after the inclusion of attribute definition like this:

```
CREATE TABLE Child(
    /* CREATE TABLE PRIMARY KEY Version 2 */
    ID      CHAR(9),
            -- Social Security Number
    Name    VARCHAR2(20),
            -- Char string up to 20 characters
    Height INT,
            -- Height in inches (Integer)
    CONSTRAINT ID_PK PRIMARY KEY(ID)
);
```

This second version of the creation of a primary key in a table is preferred for two important reasons. First, this form is the only way to designate a concatenated primary key as we saw in Chapter 1 with the **Qualifications** table. Second, we named the primary key, *ID_PK*. At times it becomes necessary to use the name of the key in an ALTER TABLE [ENABLE|DISABLE|DROP] statement. If the programmer does not assign a name, then Oracle will; however, the assigned name will not be obvious and would have to be found in a **Dictionary** table.

The third way to define a PRIMARY KEY would be to use an ALTER TABLE command. The command would look like this:

ALTER TABLE Child
    ADD **CONSTRAINT ID_PK PRIMARY KEY(ID);**
/* PRIMARY KEY definition way 3 */

As previously mentioned, this after-the-fact naming of a PRIMARY KEY tends to show a lack of planning, but it could be done this way.

As we provide more constraints, the constraint type will be appended to the chosen name. Here, the appendage is "_PK" indicating a primary key constraint. And, since the *ID* attribute is the primary key, we named the constraint "*ID_PK*."

2.1.1.2. Disallowing null Values – A CHECK CONSTRAINT

Null values present a problem in database because of what they might represent. Does a *null* symbolize:

(a) Data that is simply not knowable – perhaps a future retirement date?
(b) Data that is known, but missing? Why is it missing? If, for example, the *null* occurred in a birthday attribute, is it really unknown or does the person refuse to allow the birthday to be known.
(c) Data that is known, but does not apply for a situation?

If the table creator wanted to disallow *null* values, the way to do it would be to add the appendage "NOT NULL" to the attribute definition. Here is an example of a named NOT NULL constraint:

```
CREATE TABLE Child(
    /* CREATE TABLE PRIMARY KEY Version 2 */
    ID       CHAR(9),
             -- Social Security Number
    Name     VARCHAR2(20) CONSTRAINT
                    Name_NN NOT NULL,
             -- Up to 20 characters
    Height INT,
             -- Height in inches
    CONSTRAINT ID_PK PRIMARY KEY(ID)
);
```

We named the NOT NULL constraint as before because the situation may arise where the table may need to be altered to disable, drop, or re-enable the constraint using an ALTER TABLE command.

The primary key can never be *null* so it is unnecessary to add NOT NULL to the *ID* attribute.

2.1.1.3. CHECK Constraints on Domains

There are two types of CHECK constraints -- table constraints and attribute constraints [2.1]. If the CHECK constraint applies to a single attribute, it is specified after the attribute name and datatype in the CREATE TABLE command. A table constraint may refer to one or more attributes which are named in the constraint definition.

We shall start with an attribute constraint. Such a constraint limits the allowable values assignable to an attribute (its domain of values). The syntax for such a CHECK constraint is put in the attribute definition after the datatype as in this example:

```
CREATE TABLE Child(
    /* CREATE TABLE PRIMARY KEY Version 2 */
    ID       CHAR(9),
             -- Social Security Number
    Name     VARCHAR2(20) CONSTRAINT Name_NN NOT NULL,
```

```
            -- Up to 20 characters
    Height INT  CHECK (Height >= 24),
            -- Height in inches
    CONSTRAINT ID_PK PRIMARY KEY(ID)
);
```

In this table definition, a constraint on *Height* values has been added. No *Height* value can be assigned unless the value is greater than or equal to 24 (inches). This example as written would allow Oracle to choose the name of the constraint. In case an ALTER TABLE command should ever be required, the constraint name should be defined like this:

ALTER TABLE child
ADD CONSTRAINT Height_ck CHECK(height>=24);

In the case of *Height*, above, if for some reason, a person were permitted to enter data into the **Child** table with a *Height* violating the CHECK constraint, the following ALTER TABLE command could be used:

ALTER TABLE Child
DISABLE CONSTRAINT Height_ck;

Similar ALTER TABLE options include enabling a disabled constraint or dropping the constraint altogether with statements such as these:

ALTER TABLE Child
ENABLE CONSTRAINT Height_ck;

ALTER TABLE Child
DROP CONSTRAINT Height_ck;

When we started this section, we said there were two kinds of CHECK constraints. We illustrated a column constraint by placing restrictions on what could be entered into a column -- what values were allowable for an attribute to accept. Since the constraint was entered after the attribute definition, the column name was not included in the constraint definition.

2.1.1.4. CHECK CONSTRAINTs on More Than One Attribute

Table-level CHECK constraints include naming attributes to be constrained. Also, a CHECK constraint can be written on more than one attribute. Consider this example:

```
CREATE TABLE Reportcard(
/* Grade is an intersection attribute between a Student and a Course
    table
To assign a Grade one must reference the keys of Student and Course
The keys are presumed to be Student_id and Course_id
The referential integrity constraints are left as Exercise 2.2 */
    Student_id           VARCHAR2(30),
    Course_id            VARCHAR2(10),
    Grade_assigned       CHAR(1),
CONSTRAINT Grade_PK PRIMARY KEY (Student_id, Course_id)

/* Since two attributes are involved in this constraint, the attributes are
    named, making this a table-level constraint */
);
```

Here is another example:

```
CREATE TABLE Emp(
    Emp_no            DECIMAL(6,0)    ,
    EName             VARCHAR2(20)    NOT NULL,
    Salary            DECIMAL(8,2)    NOT NULL,
    Raise             DECIMAL(3,2),   -- Percent raise %
    CONSTRAINT Emp_no_PK PRIMARY KEY(Emp_no),
    CONSTRAINT Raise_Limit CHECK (Raise <= 0.05*Salary)
        /* A CONSTRAINT involving two attributes */
);
```

2.1.1.5. The UNIQUE CONSTRAINT

The UNIQUE constraint requires values of an attribute be matchless. This constraint is for attributes not having been designated as a PRIMARY KEY. A CREATE TABLE with a UNIQUE constraint example follows:

```
CREATE TABLE Medicine(
    Drugname      VARCHAR2(30),        --Brand name
    Chemical      VARCHAR2(30),        --Real chemical name
    Doc      VARCHAR2(30)   NOT NULL,
    CONSTRAINT Che_UK    UNIQUE (Chemical),
    CONSTRAINT Drug_PK   PRIMARY KEY (Drugname)
);
```

There are three differences between a UNIQUE and PRIMARY KEY constraint:

(a) A table may have more than one UNIQUE constraint. Tables can have only one PRIMARY KEY.
(b) A value in a UNIQUE constraint can be *null*. A PRIMARY KEY can never be *null*.
(c) UNIQUE constraints can exist in addition to the PRIMARY KEY.

2.1.1.6. A Referential Integrity CONSTRAINT

Relationships in relational databases are links between tables. Relationships are realized via foreign key/primary key constraints. Not only do we make logical connections by defining these constraints, but we also enforce referential integrity rules. A Referential Integrity **CONSTRAINT** is one in which a row in a table cannot exist if a value in that row refers to a non-existent value in another (foreign) table. To clarify the idea of referential integrity, suppose we have the following two tables representing baseball fans and their favorite team:

Teams

Team_number	TeamName
1	Pirates
2	Giants
3	Cubs
4	Phillies

Fans

Fnum	Fname	Tnum
100	Anne	2
106	Genevieve	1
102	Beryl	1
104	Mary Fran	3
108	Mary Jo	2
109	David	4
110	Johnny	
113	Rich	3

Team_number is the primary key of the **Teams** table, and *Tnum* in the **Fans** table is referred to as a foreign key. A foreign key references a primary key attribute usually found in a different table. The relationship between the **Teams** and **Fans** is through the primary key of the **Teams** table (*Team_number*), and the foreign key of the **Fans** table (*Tnum*).

To violate referential integrity, it would be inappropriate to enter a row in the **Fans** table where the *Team_number* was not defined in the **Teams** table. To try to insert the row:

<105,'Chloe',4>

in the **Fans** table would be a violation of the integrity of the database because *Team_number* 4 is not defined.

Likewise, it would be invalid to try to change or delete a value of *Tnum* in an existing row of **Fans** to make it a non-existent value in **Fans**. If, for example, in **Fans** we tried to change

<100,'Anne',2>

to

<100,'Anne',6>,

it would be wrong because there was no *Team_number* = 6.

While *Team_number* in the **Teams** table cannot be *null* due to "The Entity Integrity Constraint" (which is another way of saying primary keys cannot be *null*), *Tnum* in the **Fans** table can be *null* as it is not a primary key. It is acceptable for *Tnum* to be undefined or unknown, but it is not acceptable for *Tnum* to point to a non-existent *Team_number*. If the **Fans** table were created with *Tnum* as NOT NULL, this would constitute a "mandatory relationship." If *Tnum* can be *null*, then the relationship is an "optional relationship."

It would be invalid to DELETE a row in the **Teams** table containing a value for a **Teams** number previously defined in the **Fans** table. For example, if the following row were deleted from the **Teams** table,

<2,'Giants'>

then the row

<100,'Anne',2>

would refer to a non-existent team. It would, therefore, be a relationship with no integrity.

When this primary key/foreign key constraint is defined, we have defined the relationship of **Fans** to **Teams**.

To avoid anomalies (INSERT, UPDATE, and DELETE), there must to be a referential integrity constraint on the *Tnum* column in the **Fans** table referencing *Team_number* in the **Teams** table.

In the INSERT and UPDATE cases, we would expect (correctly) the usual action of the system would be to deny the action violating integrity. In the case of the DELETE, there are options allowing us to either disallow (RESTRICT) the DELETE (the default), CASCADE the deletion of data, or set the offended foreign key to *null* (SET NULL).

To enable a referential integrity constraint, it is necessary for the column being referenced to be first defined as a PRIMARY KEY. In the **Fans/Teams** example above, we have to first create the **Teams** table with a PRIMARY KEY. The creation statement for the **Teams** table (the referenced table) could look like this:

```
CREATE TABLE Teams
    (Team_number  NUMBER(3),
-- a CHECK on domain could be added
    TeamName     VARCHAR2(20),
        -- a CHECK could also be included here
    CONSTRAINT Team_number_pk
        PRIMARY KEY(Team_number));
```

The **Fans** table (the referencing table) would then be created using the following statement:

```
CREATE TABLE Fans
    (Fnum  NUMBER(4),
    Fname VARCHAR2(20) NOT NULL,
    Tnum  NUMBER(3),   -- NULL ALOWABLE
    CONSTRAINT Fnum_pk PRIMARY KEY (Fnum),
    CONSTRAINT Tnum_fk
    FOREIGN KEY(Tnum) REFERENCES Teams(Team_number));
```

The CREATE TABLE Fans... statement defines a column, *Tnum*, to be of datatype NUMBER(3). The statement also defines *Tnum* to be a foreign key referencing **Teams**. Within the **Teams** table, the referenced column *Team_number* must be an already defined as PRIMARY KEY.

The CREATE TABLE Teams ... must be executed and populated first. If we use CREATE TABLE as illustrated for the **Fans** table before the **Teams** table was created, we would be attempting to reference a non-existent team. Likewise, if a *Team_number* value did not exist in the **Teams** table, then referencing data could not be added to the **Fans** table.

2.1.2. Actions on CONSTRAINTS

A benefit of naming constraints explicitly is there are ALTER TABLE commands allowing a constraint to be disabled, enabled, or dropped. There may be times when a constraint may be too restrictive and needs to be lifted for some reason.

It is also possible to drop a constraint altogether. Here are some example commands:

```
ALTER TABLE Employee
DROP CONSTRAINT Raise_limit;

ALTER TABLE Child        -- temporarily disable the NOT NULL
DISABLE CONSTRAINT Name_NN;

ALTER TABLE Child        -- re-enable the NOT NULL
ENABLE CONSTRAINT Name_NN;
```

Exercises for Chapter 2

Ex 2.1.
Above we saw a table about children in kindergarten containing constraints. Create the table in your account and populate it with the inserts below. As you process additions, report and explain the results.

Add a child with values ID = 111111111, name = 'Alpha', Height = 34.
Add a child with values ID = 222222222, name = 'Beta', Height = 45.
Add a child with values ID = 111111111, name = 'Charlie', Height = 41.
Add a child with values ID = 333333333, name = null, Height = 41.
Add a child with values ID = 123123123, name = 'Donna', Height = 23.

Display the table as it now exists.

Ex 2.2.

In the chapter we created a table called **Reportcard** with attributes *Student_id*, *Course_id*, and *Grade_assigned*. We used variable character datatypes for *Student_id* and *Course_id* and a datatype of CHAR(1) for *Grade_assigned*.

Add a named CHECK CONSTRAINT on *Grade_assigned* so a letter grade must be in the set ('A','B','C','D','F').

Add a referential constraint on the **Reportcard** table to insure rows in the **Reportcard** table are linked to an existing student.

Create a **Course** table with attributes *Course_id* and *Course_name*. Populate the table with three rows.

Add a referential constraint on the **Reportcard** table to ensure rows in the **Reportcard** table are linked to an existing course.

Ex 2.3.1.
Create two tables -- **Vet** and **Dog**.

In the **Vet** table, include a *Vet _ID* for a primary key and add *Vet_address, Vet_phone,* and *Vet_city*. Make the Vet_id's 100, 200, and 300.

In the **Dog** table, use attributes *Gender, Neutered, Year_born, Dog_name, Owner,* and a foreign key, *Vett,* referencing the **Vet** table. Add and name the following constraints in **Dog**: Gender should be F or M, Neutered Y or N, Year_born no sooner than 1995. Name the referential constraint Vet_ID_FK.

Populate both tables with five rows in **Dog** and three in **Vet**. Populate the **Vet** table first (Why?). Be sure to reference all vets at least once.

Before you progress further, create a backup of the populated **Vet** and **Dog** tables you just created. To create the backup, use a command like this:

CREATE TABLE Dog_bak AS SELECT * FROM Dog;

As you do this exercise, execute INSERT commands that will and will not cause errors. Insert a dog named Fido to use Vet 200. Insert a dog named Fluffy that wants to use Vet 400. Explain why the commands worked or didn't work.

After you have finished INSERT commands, try UPDATE and DELETE commands. Update the **Dog** table changing a vet on some dog to another vet. Also, try to change a vet to a non-existing vet. Delete Vet 300 and verify the result in the **Dog** table or explain whatever error occurs.

Ex 2.3.2.

We have three vets and their Vet_id's are 100, 200, and 300.

Restore the original **Vet** and **Dog** tables from the backups. Then use ALTER TABLE to set the DELETE paradigm on the referential constraint on all dogs referencing Vets to SET NULL. Then DELETE a vet and check the **Dog** table to see if indeed the reference to the deleted vet got set to *null*.

Ex 2.3.3.

Restore all tables to the originals. Create another table called **Owner** with the attributes *Owner_name, Owner_id, Owner_address, Owner_phone*. In the CREATE TABLE, make *Owner_id* the PRIMARY KEY and add at least one constraint to each of the other attributes. INSERT at least one owner for each dog.

Now, revisit **Dog** and change the DELETE paradigm to CASCADE for the referential constraint in **Dog** referencing the **Vet** table. This time DELETE a vet and determine whether the dog referencing that vet was deleted. Then, display the **Owner** table and determine whether the owner of the deleted dog is still there or not. Explain your results.

REFERENCES

[2.1] The difference between column level and table level CHECK constraints is column level constraints apply to only one attribute in the table (a single column). Column level constraints do not specify a column name because the constraint is given when the attribute is defined and written after the column definition.

Example of a column level constraint:

Hrs_wkd   Decimal (6,2)   NOT NULL

Table level constraints name specify constrained columns.

Example of a table level constraint:

CONSTRAINT (Student_num, Class) StuClas_PK PRIMARY KEY.

Also See: https://www.quora.com/What-is-table-level-constraint-in-Oracle

Many examples of CONSTRAINTs may be found here: https://docs.oracle.com/javadb/10.8.3.0/ref/rrefsqlj13590.html

Some database examples taken from Earp, R. W., and Bagui, S. S. (2021), *A Practical Guide to Using SQL in Oracle*, 3rd Edition.

Chapter 3

THE DATABASE

3.1. INTRODUCTION

We will assume a company named Hardware_City exists. The Chief Information Officer (CIO) of the company wants a database created to keep track of customers, products, and vendors. Someone administers all computer activity for the CIO -- the System Administrator. The System Administrator deals with more than databases because there will be language programmers, security people, and others who do not directly impact the database.

Under the purview of the System Administrator, we will distinguish between "database programmers" and "language programmers." A database programmer normally deals with SQL -- querying and managing the data in the database. A language programmer might use a computer language like C++ to write programs and interfaces for end users. The language programmers may indeed use the database, but their level of involvement in the database itself is much like an end user -- they may access the data but not directly manage it. The database programmers manage data. By managing data, we mean inserting, updating, and deleting data, changing privileges, defining users, and writing scripts and procedures to access the data in SQL-like or PL/SQL language.

The System Administrator has hired several people to work on the database creation project. For database related matters, the System Administrator directly supervises the Database Administrator or DBA.

For the purpose of this work, we will assume only one DBA named Danny (Level 1) and several developmental users at lower levels in the organization chart -- Levels 2 and 3. We will define Chris, Pat, and Van as database people at Level 2, and Sam, Morgan, and Kelly as user-interface people at Level 3.

The names we chose for the people are not totally random, but rather the first letter of the name (at Levels 1 and 2) corresponds to an area of responsibility -- Danny is the DBA, Pat deals with Products, Chris with Customers, and Van with Vendors. Sam, Morgan, and Kelly are cross-area users. For example, if Sam developed queries for Sales, then Sam might need data on Vendors, Customers, and Products. Therefore, Level 3 users might write SQL retrieving data from all three areas of the database. Retrieving data involves granting access to the data one needs. Managing data at Levels 2 and 3 involves much more than just accessing the data. Managing data involves who accesses the data and when, what they are allowed to see and not, and what they are allowed to change and not. This book is more about Levels 2 and 3 rather than Level 4. Level 4 users will use normal SQL queries that do not change the database content.

Here are our levels and people in the database hierarchy we will use:

Level 0 - The System Administrator
Level 1 - Danny our DBA
Level 2 - The caretakers for areas of the database:
Chris is in charge of the **Customer** part of the database.
Pat will handle the **Product** information.
Van will manage the **Vendor** tables.
Level 3 – The developers/programmers will develop queries for non-computer people: Sam, Morgan, and Kelly -- each of whom may be accessing any part of the database to develop queries for Level 4.
Level 4 - People in various departments will use the data for day-to-

day activity: Salespeople, Accountants, etc. These people use the queries developed at Level 3.

Danny has to manage the creation of the database with the Level 2 programmers. Our original database will contain only three tables: **Customer**, **Product**, and **Vendor**. Later, there may be more objects created within each problem area, but for now, we will start with just these three tables.

Data in databases is designed to be shared. So far, we have discussed design and integrity in an effort to distinguish a good database from a poorly designed one. In this chapter, we move from the one-user, one-account mode to a multi-user environment. In consideration of other users, there must be rules governing things such as user space and how users will interact. Who creates tables? Where are the tables located? Who accesses what data? Who changes the content of the tables with INSERTs, UPDATEs, and DELETEs? With the shared-file concept comes the idea of *privileges* for each user and a defined tablespace for the database.

If we were building a house, our first considerations would be how the proposed house was designed and where the house would be located. The design affects the location and vice-versa. We should be confident about the design of the data itself, but now we must look at two issues -- where will we put the database (a tablespace) and how will we handle other users (privileges)?

The approach we will take is to first handle user interactions (privileges) and then consider the tablespace issue. The reason for this approach is that one person should be responsible for both users and space. That person is the Database Administrator -- the DBA.

3.1.1. What Is a Privilege?

A database privilege defines what a person using the database is allowed to do. In the broadest sense, there are two kinds of privileges -- ***object*** and ***system*** privileges. An object is a piece of the database itself like a table, an index, or a synonym; hence, an object privilege describes what one is allowed to do with an object. System privileges cover all privileges not dealing with objects such as the privilege to connect to the database or to define a user.

This study will involve people in a small company and the roles they play -- sort of like a stage play. We will give the people names whose first letter somewhat reflects their responsibilities. We will begin with the "computer people." However, when we say "people sharing data," we mean all people in the company in whatever department they find themselves whether they are computer people or not. To discuss sharing and a multi-user environment, we must first define some more terms and then discuss accounts and privileges usually meted out to users.

After we set up a small database and look at the problems encountered at various levels of responsibility, we will then suggest exercises to be performed by two or more people (or, less desirable, one person acting as multiple people). We will take a journey with the users and see some of things needed for them to do their job.

What privileges a person possesses (or, more properly, is "granted") depends on what the person needs to do with the data in the database -- a "need to know" situation. So, how are users and privileges managed?

3.1.2. Users, Responsibilities and Security

If everyone were going to share many files, someone must administer who-does-what with the database. At the highest level within the database, this individual is known as the database administrator (DBA). The DBA oversees the creation of the database and takes care of who-can-do-what within it -- the DBA manages privileges (among other things).

The term "user" simply means anyone who will use the database. To be sure, there is a hierarchy of users -- the DBA at the top, those just under the DBA who will care for some area of the database, and other individuals who are either end users or programmers working within this hierarchy. For our examples, we will call the DBA Level 1 and those just under the DBA Level

2. Level 2 users will be assigned a general area like customers or vendors. The next lowest layer of database responsibility is Level 3, whom we will call Developers. They will use the database to develop queries for those in Level 4 -- the end users (those not in the computer group such as accountants, purchasing agents, salespeople, etc.)

Levels of Database Users:

(1) The DBA (Database Administrator)
(2) Database managers for specific areas of the company
(3) Departmental database programmers managing specific areas of the database
(4) End users who need data to do their job, e.g., salesperson, storeroom clerk.

Databases and users' actions performed on the database must be controlled, contained, and monitored. A basic tenet of privileges and accesses is users should be granted only enough privilege and access to do their job. DBA Danny needs broad privileges. The users Danny will create will need a narrower set of privileges. By paying attention to the level of the user and granting only such privileges as necessary, the integrity and security of the database should be insured.

3.1.3. Users and Privileges

As we define people in this business example, we must define what they are allowed to do (and not do) within the database. There are two types of privileges -- object and system.

Database *objects* are things we use within the database, e.g., tables, indexes, synonyms. An object privilege defines what a person can do with an object. For example, if the object is a table, operations may include INSERT INTO or UPDATE data in the object, the table.

System privileges are general non-object actions a user may perform. An example of a system privilege would be CREATE SESSION to allow a user to connect to the database. Users need the CREATE SESSION privilege to "sign on" (connect) to the database and to manipulate or query objects within the database. There are 80 system privileges defined in Oracle 8. See reference [3.1] and other footnotes at the end of the chapter.

Commands starting with CREATE, ALTER, or DROP involve system privileges. The command, CREATE TABLE, mentions an object, a table, but creating the table is actually a system privilege.

Disclaimer: We do not portend this work is a complete coverage of privileges, but is rather an introduction by example of how the Oracle database system may evolve within a multi-user environment. Readers are encouraged to do research on each subject we cover.

Danny has the task of assigning subject areas to developers -- **Customer**s to Chris, **Product**s to Pat, and **Vendor**s to Van. Danny first needs to create the user accounts for Chris, Pat, and Van so the main tables can be created. Then, Chris needs to create the **Customer** table, and the others need to create their main tables as well. Before we discuss how Danny's account (Danny1) is created, we could begin with a basic CREATE USER command looking like this:

CREATE USER *username* IDENTIFIED BY *password*;

Example: CREATE USER Temp2 IDENTIFIED BY Temp;

To illustrate some points, we will need temporary users. We will employ Temp$n$ as a temporary user to illustrate a point. We will delete Temp$n$'s account when we are finished with this chapter. In the example above Temp2 is a user name and Temp is a password. Temp2 means this temporary user will be at Level 2 in our hierarchy of users.

3.1.4. GRANTing Privileges

The GRANT statement gives the grantee a privilege. The simplest GRANT statement has this form:

GRANT *privilege* TO *user*

Example: GRANT CREATE SESSION TO Temp2;

CREATE SESSION is a system privilege allowing Temp2 to connect to the database. Here is the result of the execution of two statements -- creating Temp2's account and granting the CREATE SESSION privilege (executed from the account of the DBA):

SQL> CREATE USER Temp2 IDENTIFIED BY Temp;

User created.

SQL> GRANT CREATE SESSION TO Temp2;

Grant succeeded.

The above two statements seem to work well, but there is a problem. While both of the above statements will execute correctly, they fall short of being useful for our database for Hardware_City.

All SQL statements have a simple form assuming defaults and a fuller form overriding some defaults. The fuller form contains embellishments to clarify the SQL. For example, SELECT can be as simple as:

SELECT * FROM Customer;

Embellishments to the SELECT are necessary to retrieve specific information. Hence, the simple SELECT usually becomes one with WHERE, ORDER BY, HAVING, nesting of SELECTs within SELECTs, and other embellishments.

Danny as DBA has a special named privilege containing all the privileges needed to deal with users and their tables. Danny has the DBA privilege which is powerful and needed by Danny alone. The DBA privilege contains the privileges Danny will need to pass to each user.

The "simple" CREATE USER above actually works. So what needs to be added for Temp2 to do work? There are three immediate problems calling for an enhancement:

(1) Temp2 can connect to the database but can't do anything else. At the very least, Temp2 has to be able to create or access tables. Danny needs to GRANT the CREATE TABLE privilege to Temp2; and so, Danny would execute this statement:

GRANT CREATE TABLE TO Temp2;

(2) After execution of the GRANT to allow Temp2 to create tables, Danny needs to consider what else Temp2 has to do. If we are going to have a hierarchy of users, user Temp2 would need to be able to create a user like Temp3 at Level 3. If we wanted Temp2 to be able to create another user, Temp2 must have the CREATE USER privilege just as Danny1 would have had to create Temp2.

At this point, you might think, "Why not let Danny deal with granting all these privileges to Level 3 users and possibly beyond?" The answer is we want a hierarchy of users; and it will be Temp2's job to handle the privileges for Temp3 at Level 3. It will be Temp3's job to manage Level 4 privileges.

While Temp2 has the privilege to connect to the database with CREATE SESSION as it stands, that privilege cannot be passed on to another user at Level 3 the way it is written above. The more correct statement for Danny to execute would be:

GRANT CREATE TABLE TO Temp2 WITH ADMIN OPTION;

3.4.1. WITH ADMIN OPTION

In GRANT statements, the appendage WITH ADMIN OPTION renders the granted privilege passable to another user. Temp2 will need the additional privileges CREATE USER and CREATE TABLE for now. Without WITH ADMIN OPTION, as above, Temp2 cannot pass along

CREATE SESSION or CREATE TABLE to the next user. Hence, Danny executes the following statements:

GRANT CREATE SESSION to Temp2 WITH ADMIN OPTION;
GRANT CREATE TABLE to Temp2 WITH ADMIN OPTION;
GRANT CREATE USER to Temp2 WITH ADMIN OPTION;

Now we test the privilege hierarchy. In moving from user to user, the CONNECT command is used. CONNECT may be abbreviated with CONN:

CONN Temp2/Temp

SQL> CREATE USER Temp3 IDENTIFIED BY Temp;

User created.

SQL> GRANT CREATE USER to Temp3;

Grant succeeded.

SQL> GRANT CREATE SESSION to Temp3;

Grant succeeded.

SQL> CONN Temp3/Temp

Connected.

SQL> CREATE USER Temp4 IDENTIFIED BY Temp;

User created.

GRANT commands can be reversed (un-GRANTed) by using the command REVOKE. For example, if Danny GRANTED CREATE USER TO Temp3, the privilege could be reversed with

REVOKE CREATE USER FROM Temp3;

3.1.5. Defining a Tablespace and Getting Started with the Database

At this point, we have seen how the DBA plans a hierarchy of users. Before actually dealing with Danny, Chris, Van, and others, we will create a space in which to store data and get started creating users, tables, and perhaps other objects.

Let us begin our database construction by creating the account for Danny1 with a few more embellishments to the above examples. First of all, we would like to create a special place to store the data for Hardware_City. We want to define a space to hold our database data; this space is defined as a tablespace. If a tablespace is not created specifically for Danny and the other users under Danny, then they would use the default tablespace USERS. It will prove best if we contain the data in one named tablespace because there are SQL commands at the tablespace level, and we can use tablespace manipulation to deal with just the users in Hardware City.

The CREATE TABLESPACE command precedes the CREATE USER Danny1. It would look like this:

CREATE TABLESPACE HARDWARE_CITY
/* This command is executed by the System Administrator who will next create Danny1 */
DATAFILE 'hardware_city.dbf'

The name of the tablespace will be *HARDWARE_CITY*. The name of the actual file on the host computer is *hardware_city.dbf*.

3.5.1. The DBA Is Created by the System Administrator

Now that we know where we will put the database, the System Administrator can create Danny's account including the named TABLESPACE:

CREATE USER Danny1 IDENTIFIED BY Danny;
DEFAULT TABLESPACE HARDWARE_CITY;
QUOTA 20M ON HARDWARE_CITY;

In this command, the *username* is Danny1 and the *password* is Danny. We have included in the CREATE USER command the defined tablespace, *HARDWARE_CITY*, together with a QUOTA to limit the amount of data included in the tablespace by Danny1 (20 megabytes).

Since the user, Danny1, has been created, some privileges must be granted by the System Administrator to Danny1. In database terminology, a "ROLE" is a collection of privileges. As previously mentioned, the privileges assigned to Danny1 come from a built-in, powerful ROLE called the DBA ROLE. Roles can be user-defined, and we will do this presently. However, for now, we will have Danny1 assigned the built-in ROLE of DBA. This DBA ROLE is assigned to Danny1 by the System Administrator who created the account for Danny1:

GRANT DBA to Danny1;

The proper sequence of events so far is:

(1) The tablespace *HARDWARE_CITY* is created and will be our default tablespace.
(2) Danny1 is created by the System Administrator using our default tablespace.
(3) Danny1 is granted the DBA ROLE by the System Administrator.
(4) Danny1 will now create and assign privileges to Pat2, Chris2, and Van2 because the DBA ROLE allows Danny to do so.

In this scenario, we are going to have only one DBA, Danny, whose *username* and *password* are Danny1/Danny.

3.5.2. Level 2 Users Are Created by the DBA

With Danny1's account created with DBA privileges; we want to begin developing our database with Level 2 users to create our main tables. We need to plan what privileges Danny might GRANT to each user. The DBA assigns privileges to users who may, if they have the administrative power, re-assign privileges to other users. As we saw above in the case of Temp2, the GRANT command must contain WITH ADMIN OPTION for privileges to be passed through to another user.

To begin, Danny1 connects to the database and a *session* begins. A session means a person has signed on (connected with the database). The signed-on status will remain in effect until the person signs off or disconnects by connecting as someone else.

Danny will create an account for Pat and then grant some privileges. Pat will need to connect to the database (the CREATE SESSION privilege) and create a **Product** table (using the CREATE TABLE privilege). Furthermore, Pat will need CREATE USER to create an account for the Level 3 person, Morgan.

Danny connects to Oracle then executes this series of commands:

```
CREATE USER Pat2 IDENTIFIED BY Pat;
DEFAULT TABLESPACE HARDWARE_CITY;
QUOTA 2M ON HARDWARE_CITY;
```

For now, Danny allocates 2 megabytes for Pat to use. (This space can be enhanced later should the need arise.) Since the account Pat2 is now created, Danny begins assigning privileges to Pat with statements like we saw before. After the GRANTs from Danny, Pat2 connects and then this Data Dictionary view-query is executed from Pat2's account. (We leave the actual GRANT command Danny1 executes to the reader as an exercise.)

```
SQL> SELECT * FROM USER_SYS_PRIVS;
```

USERNAME	PRIVILEGE	ADMIN_OPT
Pat2	CREATE SESSION	YES
Pat2	CREATE TABLE	YES
Pat2	CREATE USER	YES

Note: if you run this query, you may have to run a command to set the column width, e.g., COLUMN USERNAME FORMAT a10.

3.1.6. Using Scripts

Now that we have seen the general pattern of how the users are created, it is more efficient to use scripts to do so. A script is a series of commands stored on the host computer. We presume the host computer is using UNIX as an operating system. To create a script, the following steps are taken.

Step 1. From SQL*Plus the command, **host**, is executed.

Step 2. In the host operating system, a text file is created containing the commands a person would use in SQL. The text file must have the sql file type appended. So, if a text file is created on the host and called "Create_User," the text file is stored in UNIX as Create_User.sql.

Step 3. We issue the command, exit, in the host to exit back to SQL*Plus. Notice that the command, exit, is lower case. UNIX is extremely case sensitive.

Step 4. We execute the script using the format:

@Create_user

You might wonder why we'd go to such trouble to execute a command this way -- exit to the host, create a text file, exit from the host, and execute the script. The answer is flexibility. Once we have a script created, it is easier on the person running a command repeatedly to use scripts.

Example:

Danny could simply issue the command:

CREATE USER Pat2 IDENTIFIED BY Pat;

or

Danny could exit to the host and create a text file called Create_user. The text file might look like this:

/* Create_user.sql written by Danny July 4, 2021
 This script will be used to create a user named below */
CREATE USER Pat2 IDENTIFIED BY Pat;

Danny stores the script, exits back to SQL, and then executes from within SQL:

@Create_User

The result of a command issued and a script executed is the same. The difference is that in the script, Danny can embellish the CREATE USER command and can use the script for Van and Chris by changing the name and password in the CREATE USER command.

As we mentioned above, the actual CREATE USER command is a bit more complicated than just CREATE USER .. IDENTIFIED BY .. Instead, the command needs a defined tablespace

The script is altered to include the tablespace, so it now looks like:

/* Create_user.sql written by Danny July 4, 2021
This script will be used to create a user named below */
CREATE USER Pat2 IDENTIFIED BY Pat;
DEFAULT TABLESPACE HARDWARE_CITY;
QUOTA 2M ON HARDWARE_CITY;

And, to create user Van2 and Chris2, Danny needs only to change Pat2 to Van2 and Pat to Van and re-run the script rather than writing three lines of code for each user. As queries get more complicated, it is far easier to change a script than to re-type commands.

We will deal with the actual creation of users with privileges in the Exercises at the end of the chapter. Then we will consider privileges and table creation in Chapter 4.

3.1.7. Setting the Environment with a Sign-on Script

It is good to set up SQL accounts with a "housekeeping" script defining the environment for working with SQL. To do so, we will create a script of useful SET commands. SET commands are used to define the maximum width of the output on the screen (LINESIZE), the vertical page size (PAGESIZE), and even the default prompt (SQLPROMPT).

There are 67 SET commands defined in the online SQL*Plus® User's Guide and Reference [3.1]. We will suggest a "startup script" using only two (LINESIZE and WRAP) of the 67, but you are encouraged to explore all of them. To see all the SET possibilities, the appropriate help command is HELP SET.

We will call the startup script, Runfirst. When the user connects to SQL*Plus for the first session of the day and executes Runfirst, the environment is initialized. We will illustrate Runfirst to set the editor, LINESIZE, and WRAP.

The editor may be any UNIX editor, but we use vi.

The LINESIZE sets the maximum number of characters on the computer screen when a command is executed -- the maximum screen width.

WRAP should be set to ON. If it is set OFF and should a result have a wider output than LINESIZE, the result would be truncated. While SET WRAP ON is the default, it may be possible that some other script could have set it OFF.

SQL*Plus SET commands are not persistent; therefore, the Runfirst script should be executed with each sign on.

Just as Create_User script was created and used, Runfirst will be handled in a similar fashion. The text file on the host computer will be named "Runfirst.sql," and the script would be executed from SQL*Plus as:

SQL> @Runfirst

The text file Runfirst will contain the following code:

/* Runfirst.sql */
/* Initialize environmental parameters – Run this script upon signing on */
/* June 21 2021 */
/* Created by *yourname* */
define_editor=vi
SET LIN 100
/* LIN = linesize */
SET WRAP OFF
/* other SET commands could be added here */
PROMPT Editor is defined as vi
PROMPT Parameters LINESIZE and WRAP are SET
PROMPT Widen your window to accommodate 100 characters
/* Runfirst.sql written by Danny July 5, 2021
The script is to be executed upon sign on. The SET commands
Found herein are not persistent */

Exercises for Chapter 3

Ex 3.1.
Have the System Administrator create the tablespace for the company and the DBA, Danny1/Danny. It is common practice to refer to accounts in this *username/password* format.

Ex 3.2.

As Danny1, create a script in a host text file called Readme.sql. Have the script contain the following:

/* My first script *your name*, the *date* */
SHOW USER

Exit the host back to SQL; execute the script.

Ex 3.3.
From Danny1, run two commands:

DESC Dict

and

SELECT * FROM Dict;

DESC is a synonym for the command, DESCRIBE. Dict is a synonym for the table, Dictionary. The DESC command shows us the structure of the table, Dictionary. Dictionary is an index of over 2,000 rows of dictionary table names that are maintained by Oracle. The result of the SELECT will scroll by quickly and will look quite disorganized. Execute the command HELP SET. This command will show all the SET commands used to set up the environment for a user. Execute the command SET LINESIZE = 50 and rerun the SELECT * FROM DICT command. Then, execute the command SET PAUSE ON and execute the SELECT again. To make the SELECT command stop selecting, use the <Esc> key.

Execute the command, SET ROWNUM 5.

Repeat the SELECT from above first by setting the LINESIZE to 50 and use SET WRAP ON. Then, use SET WRAP OFF and repeat the SELECT.

The point of this exercise is to set environmental values for many things, such as LINESIZE and WRAP. SET other parameters you find in the HELP SET output and see what changes (if anything) in your command SELECT * FROM Dict.

Important – When this exercise is completed, execute the command SET ROWNUM 0 to reset row counting.

Ex 3.4.

Log on as Danny1 and create the scripts Create_user and Runfirst from Danny1.

(1) Execute Runfirst.
(2) Execute Create_user creating Pat2.
(3) Modify the Create_user script to create Van2 and run it again.

REFERENCES

[3.1] http://docs.oracle.com/cd/A64702_01/doc/server.805/a58397/ch21.htm Oracle8 Administrator's Guide, Release 8.0, A58397-01, Chapter 2.
[3.2] http://www.dba-oracle.com/t_sql_plus_column_format.htm
[3.3] Oracle provides excellent web support for all commands and topics related to SQL. A list of SET commands and what they do may be found at: https://docs.oracle.com/cd/E11882_01/server.112/e16604/ch_twelve040.htm#SQPUG060.

Chapter 4

PRIVILEGES AND ROLES

4.1. INTRODUCTION

The account Pat2 is created and Pat is anxious to create the **Product** table. If this were an exercise of a single user, Pat would simply issue a CREATE TABLE command, load the table with some data, and move on. However, this is a situation where Pat and the whole database crew know Pat is responsible for the **Product** table and more people than just Pat will be using **Product**. Since Pat and everybody else will be sharing their data, we have to deal with granting privileges to each person. Who can access what data and how can their access be controlled?

In Chapter 3, privileges were allotted to Pat by Danny, one at a time. No other users were granted privileges. In this chapter, we will demonstrate how and why it is a better idea to use scripts to create users and assign privileges via ROLEs.

4.1.1. Start with a Clean Slate

Where are we with Pat? In Chapter 3, we created an account for Pat (Pat2). Pat was granted some basic privileges because Pat needs some privileges to work within the database. Pat received the privileges CREATE

SESSION, CREATE TABLE, and CREATE USER. In the creation of Pat2, Pat and the other users are all organized in a tablespace named Hardware_City.

Privileges are best handled by defining ROLEs. A ROLE is a collection of privileges that can be managed and used by the grantor to more efficiently and allocate privileges. It is efficient to deal with ROLE management using scripts. In this chapter, we begin by using what we have learned and starting over for the sake of uniformity. It is like we clean the chalk board; and knowing what we know from previous chapters, do the creations and privileges one way for all users.

Our first action will be to clear out any users under Danny. Danny issues these commands:

DROP USER Pat2 CASCADE;
DROP USER Van2 CASCADE;
DROP USER Chris2 CASCADE;
DROP USER Kelly3 CASCADE;
DROP USER Morgan3 CASCADE;
DROP USER Sam3 CASCADE;

The optional CASCADE appendage insures that any object created by any of the users is erased along with the user.

Finally, we created a temporary user, Temp2, in Chapter 3 as an example. Just in case there is any trace of that user remaining, we will erase that account as well.

DROP USER Temp2 CASCADE;

4.1.2. Create Level 2 Users

To create users at Level 2, we can use a script similar to the one in Chapter 3. In this one, we will use a PROMPT command to queue the user

Privileges and ROLEs 61

(Danny) for the name of the user we want to create. The PROMPT command will be followed by an ACCEPT command:

```
/* Create_userV2.sql CREATE USERS, Version 2, written by Danny July 4, 2021
    This script will be used to create a user named below */
PROMPT Enter a user name.
ACCEPT &user_name "User name --> " CHAR
PROMPT An account will be created for &user_name||2
PROMPT The password will be &user_name
DROP USER &user_name CASCADE;
CREATE USER &user_name||2 IDENTIFIED BY &user_name;
DEFAULT TABLESPACE HARDWARE_CITY;
QUOTA 2M ON HARDWARE_CITY;
```

The script would be run by Danny for Pat, Van, and Chris, creating their accounts with all the same space. We included another DROP command to insure there is no "debris" left from previous activity.

4.1.3. Create a ROLE for Level 2 Users

The management of privileges is also best handled through user-created ROLEs. A ROLE is a set of privileges assignable to a user. It is more efficient to manage a ROLE than to deal with each privilege for each user one at a time.

Now that Danny has created three users, Pat2, Van2, and Chris2, it is time to assign privileges to them. To do that, we create a ROLE for Level 2 and start assigning privileges to the ROLE:

```
CREATE ROLE ROLE2;
GRANT CREATE SESSION TO ROLE2 WITH ADMIN OPTION;
```

All Level 2 personnel need the CREATE SESSION privilege to connect to the database. Level 2 people will need to create accounts for Level 3 and they need to be able to pass along CREATE SESSION -- hence, the necessity for the WITH ADMIN OPTION.

Level 2 people will also need the CREATE TABLE privilege to create our three main tables, **Vendor**, **Customer**, and **Product**. The CREATE TABLE may be needed at Level 3 and again, WITH ADMIN OPTION is warranted:

GRANT CREATE TABLE TO ROLE2 WITH ADMIN OPTION;

Level 2 will need to create Level 3 users and will be granted the CREATE USER privilege. However, this GRANT has no need to pass along CREATE USER to Level 4:

GRANT CREATE USER TO ROLE2;

4.3.1. Empower Level 2 Users
Danny may now GRANT ROLE2 to the Level 2 personnel. This action requires three simple statements:

GRANT ROLE2 to Pat2;
GRANT ROLE2 to Van2;
GRANT ROLE2 to Chris2;

Had we created ROLE2 before we created users, these statements could have been part of the CREATE USER script.

4.1.4. Level 2 CREATE TABLES

We now have our users defined with the privileges they need to create their tables. The easiest way to do this is for each of the Level 2 users to create a script for table creation. The reason for using a script is because the creation of the table may have to be redone or altered. Furthermore, the CREATE TABLE command is rife with constraints. Should there be an error in the execution of the create script, the evolution to a correct script if far easier if we start with a script. Here it is:

```
/* The creation of the Product table from the account of Pat2 */
 DROP TABLE Product;
 /* Just in case of a re-creation, this command would refresh the creation.
When the script is run for the first time, it will give an error and the first-
time error may be ignored. */
 CREATE TABLE Product
 (Product_ID NUMBER(4),
 Pname      VARCHAR(20)         NOT NULL,
  Ptype     VARCHAR(20)         NOT NULL,/*Type of Product */
 QOH        INTEGER,            /*Quantity on hand */
 Price      NUMBER(7,2),        /* Normal selling price */
 Item_type  VARCHAR(17)         NOT NULL,
 /* How sold (by the item, by the package ..) */
 CONSTRAINT Product_PK PRIMARY KEY(Product_ID)
  );
 /* stored as Create_product.sql on host */
```

After creating this text file on the host, Pat reconnects to SQL and executes the above script with @Create_product.

Users Van and Chris will write similar scripts as an Exercise.

Pat's job is now to populate the **Product** table. Here again, should **Product** need to be recreated or should there be a change in the data, a script is most appropriate for table loading. After exiting to the host, Pat creates this text file:

```
/* Load the Product table written by Pat July 4, 2021*/
INSERT INTO Product VALUES
(1000,'Saw Blades','Tools',100,9.85,'PKG');
INSERT INTO Product VALUES
```

(2000,'Paint Buckets','Paint',452,3.95,'ITEM');
INSERT INTO Product VALUES
(3000,'Sheet Metal Screws','Hardware',12500,1.45,'PKG OF 5');
INSERT INTO Product VALUES
(4000,'Wall Sockets','Electrical',124,7.68,'ITEM');
INSERT INTO Product VALUES
(5000,'Citronella','Chemicals',25,6.24,'Candle');
/* stored as Load_product.sql on host *

Since Pat has created a table and populated it, Pat can view the contents of table with a "SELECT *" statement. It is prudent to define the column headings and sizes so the output looks reasonably good. Again, a script is the best way to handle this task because if the output does not look good, the script can be easily changed. "Good" is subjective, but clearly one needs to control the look of output. Here is an example of how to do it:

/* Display the contents of the **Product** table. Written by Pat July 7, 2021 */

```
COLUMN Product_id HEADING "PID"      FORMAT 9999
COLUMN Pname      HEADING "Pname"    FORMAT a20
COLUMN PType                         FORMAT a10
COLUMN Qoh                           FORMAT 99999
COLUMN Price                         FORMAT 99999.99
COLUMN itemtype                      FORMAT a12
SQL> SELECT * FROM Product;
```

The result set:

PID	Pname	Ptype	QOH	Price	Item_type
1000	Saw Blades	Tools	100	9.85	PKG
2000	Paint Buckets	Paint	452	3.95	ITEM
3000	Sheet Metal Screws	Hardware	12500	1.45	PKG OF 5
4000	Wall Sockets	Electrical	124	7.68	ITEM
5000	Citronella	Chemicals	25	6.24	Candle

Privileges and ROLEs   65

The reason Pat could perform this query is Pat created **Product**; the terminology used is *Pat owns **Product***. A person can manipulate and manage objects they own without explicit GRANTs.

4.1.5. Level 3 Users Created

Pat now needs to create the user, Morgan. This time we will not need the PROMPT/ACCEPT from the earlier example because Pat has only one user to create. From Pat's account, we write and execute this CREATE USER script: (How is Pat allowed to create a user?)

```
/* Create a Level 3 User written by Pat July 8 2021 */
/* Filename is CreateU.sql */
DROP USER Morgan3 CASCADE;
CREATE USER Morgan3 IDENTIFIED BY Morgan;
DEFAULT TABLESPACE HARDWARE_CITY;
QUOTA 1M ON HARDWARE_CITY;
GRANT CREATE SESSION TO Morgan;
GRANT SELECT on Product TO Morgan;
```

The first command is to DROP Morgan's account entirely just in case it had even been created. We want these scripts to start anew. Since Morgan works for Pat, Pat gave Morgan the power to connect to the database and also to select data from the **Product** table owned by Pat.

To test Morgan's account, Morgan connects and does a SELECT on the **Product** table:

```
SQL> CONN Morgan3/Morgan
Connected.
1* SELECT * FROM Pat2.Product;
PID Pname Ptype QOH Price      Item_type
----- -------------------- ----------   -----   ---------- ------------
1000 Saw Blades ... same as above
```

4.1.6. CREATE SYNONYM

The table name for **Product** from Morgan3's account had to be qualified (Pat2.Product). It would be convenient for Morgan if a synonym were created for **Product**. Since the current connection to the database is via Morgan, the creation of a synonym would be:

SQL> CREATE SYNONYM Product FOR Pat2.Product;
CREATE SYNONYM Product FOR Pat2.Product
*
ERROR at line 1:
ORA-01031: insufficient privileges

Oops! The problem is Pat2 needed to GRANT the CREATE SYNONYM privilege to Morgan3. So we connect again as Pat2 and execute the GRANT:

SQL> CONN Pat2/Pat

Connected.

SQL> GRANT CREATE SYNONYM TO Morgan3;
GRANT CREATE SYNONYM TO Morgan3
*
ERROR at line 1:
ORA-01031: insufficient privileges

Oops again! So Pat does not have the CREATE SYNONYM either, so we must go back to Danny1 and have Danny GRANT the CREATE SYNONYM privilege to Pat2. Remember Danny is the DBA and the DBA ROLE contains broad privileges, one of which includes CREATE SYNONYM WITH ADMIN OPTION. Pat is GRANTed CREATE SYNONYM. Since Pat needs to pass the privilege to Morgan, the GRANT from Danny to Pat must be done with the WITH ADMIN OPTION attached:

Privileges and ROLEs 67

SQL> CONN Danny1/Danny

Connected.

SQL> GRANT CREATE SYNONYM TO Pat2 WITH ADMIN OPTION;

Grant succeeded.

But wait! Instead of Danny granting Pat this privilege, wouldn't it be better if Danny were able to deal with not only Pat, but also Van and Chris as well. Instead of granting the CREATE SYNONYM directly to Pat, what Danny could to is to alter ROLE2 to include this privilege. Danny modifies and executes the last command:

SQL> GRANT CREATE SYNONYM TO ROLE2 WITH ADMIN OPTION;

Now, not only were Pat's privileges modified but also anyone with the ROLE2 privilege now has the same privileges as Pat. Pat proceeds to deal with Morgan:

SQL> CONN Pat2/Pat

Connected.

SQL> GRANT CREATE SYNONYM TO Morgan3;

Grant succeeded.

SQL> CONN Morgan3/Morgan

Connected.

SQL> CREATE SYNONYM Product FOR Pat2.Product;

Synonym created.

SQL> SELECT * FROM Product;

PID	Pname	Ptype	QOH	Price	Item_type
1000	Saw Blades	Tools	100	9.85	PKG
2000	Paint Buckets	Paint	452	3.95	ITEM
3000	Sheet Metal Screws	Hardware	12500	1.45	PKG OF 5
4000	Wall Sockets	Electrical	124	7.68	ITEM
5000	Citronella	Chemicals	25	6.24	Candle

4.1.7. GRANTing to PUBLIC

Rather than dealing with grants to each of the Level 2 and 3 people to access a table, this command could have been used:

GRANT SELECT on *tablename* to PUBLIC;

This way, anyone can see what's in the table, *tablename*. If a table is truly common information such as a table of area codes or state abbreviations, granting access to PUBLIC might be okay; however, granting to PUBLIC is a loose way to handle security. A grant to PUBLIC allows access to anyone at any time. Also, if the information were so ubiquitous it did not need security, then it would seem odd if some one person were not allowed to see the information. Why would you GRANT SELECT access to PUBLIC and then REVOKE the privilege from someone?

We suggest you avoid the PUBLIC option for GRANTing privileges. The security using PUBLIC grants is just too loose.

Exercises for Chapter 4

Ex 4.1.
If you have not done so, DROP all users with CASCADE. Then, write and store the CREATE USER script; execute the script to create Van2, Chris2, and Pat2.

Ex 4.2.
Create ROLE2 as we did in the chapter and GRANT ROLE2 to the Level 2 users.

Ex 4.3.
Write, store, and execute the script to load the **Product** table as Pat.

Ex 4.4.
Write, store, and execute the script to display the **Product** table from Pat2.

Ex 4.5.
Create ROLE3 with the privileges CREATE SESSION, CREATE TABLE, and CREATE SYNONYM. This must be done as Pat2, Van2, and Chris2.

Ex 4.6.
From each Level 2 user, write, store, and execute a script to create the appropriate Level 3 user. You don't need PROMPT/ACCEPT as each Level 3 is answerable to a specific Level 2 user. Also, include the GRANT of ROLE3 as the last line of the scripts.

As a reminder, here is the organization chart:

Level 0 - The System Administrator
Level 1 - Danny our DBA
Level 2 – Pat
Level 3 – Morgan

Level 2 – Chris
Level 3 – Kelly
Level 2 – Van
Level 3 – Sam

Level 4 - People in various departments using the data for day-to-day activity, such as Salespeople, Accountants, etc. These people use the queries developed at Level 3.

Ex 4.7.

For each Level 2 person, write scripts to create the appropriate tables: **Vendor** by Van2 and **Customer** by Chris2.

Ex 4.8.

For each Level 2 person, write scripts to load their tables. At the end of this exercise, the tables **Product**, **Vendor**, and **Customer** should be created and populated.

Ex 4.9.

For each Level 3 person, CONNECT and create a synonym for the main tables.

Ex 4.10.

Connect as each Level 2 and 3 person and execute these commands:

SELECT COUNT(*) FROM Vendor;
SELECT COUNT(*) FROM Customer;
SELECT COUNT(*) FROM Product;

All privileges should be in place and all synonyms defined. Should there be a problem with any user executing these three commands, fix it.

Ex 4.11.

From any account, connect to the database and verify the main tables look like this:

SQL> @Vendor_query

Vid	Vname	Vaddr	Vcity	ST	Vzip	Vphone
100	Acme Supply	123 PTree St.	Atlanta	GA	30002	4045551325
200	Norris Pumps	6 Palafox St.	Pensacola	FL	32501	8505551313
300	Petro Pipes	99 University Av	Tuscaloosa	AL	35402	3345551966
400	Hardy Hardware	33 Tornillo Dr.	Los Angeles	CA	91007	6265551234
500	Snow Screws	82 Fargo Rd.	Shakopee	MN	55378	9525556789

SQL> @Product_query

Pid	Pname	Ptype	Qty on hand	Price	Type
1000	Saw Blades	Tools	100	9.85	PKG
2000	Paint Buckets	Paint	452	3.95	ITEM
3000	Sheet Metal Screws	Hardware	12500	1.45	PKG OF 5
4000	Wall Sockets	Electrical	124	7.68	ITEM
5000	Citronella	Chemicals	25	6.24	Candle

SQL> @Customer_query

Cid	Cname	Caddr	Ccity	ST	zip	Phone
330	Abbie Walker	1988 Druid Hwy.	Cumming	GA	30041	6785557272
335	Matt Houston	15823 Fish Lane.	Gainesville	GA	30508	7705550001
340	Daphne Jayne	1 Small Ct.	Dunwoody	GA	30347	6785552016
345	Ellie Texann	2014 Newly Blvd.	Brookhaven	GA	30319	4045553333
350	Penny Penn	77 Nopound St.	Atlanta	GA	30303	4045559996

Chapter 5

THE DICTIONARY

We now have set up Levels 2 and 3, the main tables, **Vendor, Customer,** and **Product**. We have roles and privileges defined. At this point, it is appropriate to step back and look at what we have and to check that everything is verifiable in the dictionary. The first task is to consider how to approach the dictionary.

5.1. THE DICTIONARY PARADIGM

In accessing tables in the Data Dictionary, here are the steps we will follow:

1. Find the table you think you want to see.
2. Use SELECT COUNT(*) FROM *table_name*.
3. Use DESC *table_name*.
4. Choose your columns and Use COLUMN formatting if appropriate.
5. SELECT * or SELECT *named columns*.

How do we know what table we want to see?

One of the first things to do in dealing with the dictionary is to look at the **Dictionary** table itself. This table is an index of all tables in the dictionary. Knowing how the dictionary is set up, we can proceed with Step 1 – finding the specific table we want to see.

What do rows in the **Dictionary** table look like? There are only two attributes:

```
SQL> DESC Dictionary
Name                                    Null?    Type
--------------------------------------  -------- ----------------------------
TABLE_NAME                                       VARCHAR2(30)
COMMENTS
```

How many rows are there in the **Dictionary** table?

```
SQL> SELECT COUNT(*) FROM DICT;  /*  DICT is a synonym for Dictionary*/
COUNT(*)
----------
2553
```

In effectively looking up information in the dictionary, we need to know which of the 2553 tables may be of interest. Suppose we want to look for information about tablespaces.

5.2. DRILLING DOWN INTO INFORMATION IN THE DICTIONARY

Step 1. Find the table you think you want to see.

```
SQL> SELECT    TABLE_NAME
  2 FROM       DICT
```

3 WHERE UPPER(TABLE_NAME) LIKE '%TABLESPACE%';

In this query, we upper case the name of the table just in case a *TABLE_NAME* is in lower or mixed case. After looking at the result of this query, we settle on a table to examine, **USER_TABLESPACES**.

Step 2. How many rows are in **USER_TABLESPACES**?

SQL> SELECT COUNT(*) FROM USER_TABLESPACES;

COUNT(*)

12

What is in these 12 rows?

Step 3. Describe the table of interest.

SQL> DESC USER_TABLESPACES

Name	Null?	Type
TABLESPACE_NAME	NOT NULL	VARCHAR2(30)
BLOCK_SIZE	NOT NULL	NUMBER
INITIAL_EXTENT		NUMBER
NEXT_EXTENT		NUMBER
MIN_EXTENTS	NOT NULL	NUMBER
MAX_EXTENTS		NUMBER
MAX_SIZE		NUMBER
PCT_INCREASE		NUMBER
MIN_EXTLEN		NUMBER
STATUS		VARCHAR2(9)
CONTENTS		VARCHAR2(9)
LOGGING		VARCHAR2(9)

```
 FORCE_LOGGING                          VARCHAR2(3)
EXTENT_MANAGEMENT VARCHAR2(10)
 ALLOCATION_TYPE                        VARCHAR2(9)
 SEGMENT_SPACE_MANAGEMENT    VARCHAR2(6)
 DEF_TAB_COMPRESSION                VARCHAR2(8)
 RETENTION                  VARCHAR2(11)
 BIGFILE                                VARCHAR2(3)
 PREDICATE_EVALUATION               VARCHAR2(7)
 ENCRYPTED                              VARCHAR2(3)
 COMPRESS_FOR              VARCHAR2(12)
```

In the dictionary, many tables are very broad like this one -- many attributes. Tables in the dictionary often consist of numerous columns with what seems to be odd information in them. This is odd in the sense you could spend a great deal of time becoming an expert on the contents of one table when most of the time you might just want to see one or two interesting columns.

Step 4. Choose columns and use COLUMN formatting if appropriate.

Here we'd like to see just the names of tablespaces, so we choose only one attribute. We can skip the formatting because table names have a maximum length of 30 characters.

Step 5. SELECT * or SELECT *named columns*

```
SQL> SELECT TABLESPACE_NAME FROM
USER_TABLESPACES;

TABLESPACE_NAME
--------------------------------------------------------------------------------
SYSTEM
SYSAUX
UNDOTBS1
```

TEMP
USERS
EXAMPLE
DBCLASS
DBCLASS3
DBTEMP
HARDWARE
TEMPH
HARDWARE_CITY <--- here we are!

12 rows selected.

Had we chosen several attributes for the result set, we would suggest using *column formatting* as we did in the earlier example.

Here is a model for column formatting:

COLUMN *column_name* HEADING "*put heading here*" FORMAT a20

(This COLUMN statement would be for a situation where the *column_name*s you want are 20 characters or less. The HEADING is optional. See section 3.2.)

The Data Dictionary has tables beginning with USER_, ALL_, and DBA_. Each user has been granted privileges. Based on the level of privileges one has, some of the **Dictionary** tables may or may not be readable. The ones beginning with USER_ are usually accessible. Most of the ones that begin with ALL_ will be readable. Less of the tables beginning with DBA_ may be accessible.

Just because you see the name of the table in the dictionary does not mean it is accessible to you. Given that, there is no harm in trying to look at any dictionary view or table for which you know the name.

Exercises for Chapter 5

Ex 5.1.

How many dictionary tables deal with privileges? Of these, you will notice some are interesting:

ROLE_ROLE_PRIVS
ROLE_SYS_PRIVS
ROLE_TAB_PRIVS
SESSION_PRIVS
TABLE_PRIVILEGES
USER_SYS_PRIVS
USER_TAB_PRIVS
USER_TAB_PRIVS_MADE
USER_TAB_PRIVS_RECD

Explore each of these using the Dictionary Paradigm.

Ex 5.2.

Look at the information in these particular tables:

ALL_TAB_PRIVS
ALL_TAB_PRIVS_MADE
ALL_TAB_PRIVS_RECD

Are all of the users at the same level of privilege? Have all the Level 2 users been GRANTed the same privileges? Level 3?

Ex 5.3.

Write a script to create and populate a table of people or something meaningful to you, for example, a list of your friends and their phone numbers. The table may be something like one of these:

Friends (Name, Phone)
Appointments (Name, Date)

Books_read (Title, Author)

You can add more information if you like. The point of this exercise is to write a script containing both the table creation as well as appropriate INSERT commands. Include no less than five rows in your table. (It does not have to be realistic.) It is suggested the script start with:

DROP TABLE Friends;
CREATE TABLE Friends ...;
INSERT INTO Friends ...;

When the script has been executed, GRANT another person SELECT privileges on your table via a user-defined ROLE called MYROLE which you will create. Then, use the Data Dictionary to view information about your table and the GRANT you made. With a WHERE clause, filter the result set to return only this one table, e.g., Friends, in each of the following dictionary tables:

USER_TABLES, ALL_TAB_PRIVS, ALL_TAB_PRIVS_MADE, ALL_TAB_PRIVS_RECD, USER_TAB_PRIVS, USER_TAB_PRIVS_MADE, USER_TAB_PRIVS_RECD, DBA_TAB_PRIVS, DBA_TAB_PRIVS_MADE, DBA_TAB_PRIVS_RECD, ROLE_TAB_PRIVS, SESSION_PRIVS, and TABLE_PRIVILEGES.

Chapter 6

ACCESSING OTHER USERS' TABLES WITH SCRIPTS

6.1. BACKUPS

In Chapter 4, we created our database staff accounts. With ROLEs we granted appropriate privileges to users. We also had all users create synonyms for the main tables **Vendor**, **Customer**, and **Product**. In this chapter, we want to expand the privileges for Level 3 users. We also want to backup our tables and audit what these users do.

Backing up the three main tables is a task for the Level 2 users. Each should execute a command like this from Pat2:

CREATE TABLE Product_bak0807 AS SELECT * FROM Product;

It is important to keep backups in a timely way. "Timely" depends on the traffic of activity on the table. If there are only a handful of DML commands affecting a main table, then perhaps a daily backup would be appropriate. If there are many changes, then backups may be hourly or even more frequently. It is impossible to state the frequency of backup creations without some idea of how often the table changes. The responsibility of backup creation and how long the backup should be kept would depend on the Level 2 person responsible for the table.

6.2. AUDITING

Now that we are dealing with multiple people accessing and changing our main tables, we need to audit the tables to verify changes before and after the fact.

An audit script might include counts, sums, averages, minimum, and maximum values. Part of such a script could look like this:

```
/* Audit Customer written by Pat 07/14/2021 */
SELECT COUNT(*) FROM Product;
SELECT COUNT(Ptype) FROM Product;
SELECT COUNT(QOH), SUM(QOH), MAX(QOH), MIN(QOH), AVG(QOH)
     FROM Product;
SELECT Item_type, SUM(QOH) FROM Product
     GROUP BY Item_type);
/* Add more auditable attributes and share your script with the class */
```

As Exercise 6.2, we will need to create audit scripts for each main user at Level 2. To refresh your memory, here are the tables and attributes of our main tables:

Vendor (<u>Vid</u>, Vname, Vaddr, Vcity, VST, Vzip, Vphone)
Product (<u>Product_ID</u>, Pname, Ptype, QOH, Price, Item_type)
Customer (<u>Cid</u>, Cname, Caddr, Ccity, CST, Czip, CPhone

6.3. GRANTING DML PRIVILEGES TO LEVEL 3 USERS

The granting of DML commands to Level 3 users should be part of the overall plan. The only catch here is to be sure the privileges for each main table are restricted to the Level 3 person for that table. Let us see what Van2 needs to do and then we will leave Pat2 and Chris2 to deal with their Level 3s. Here is a reminder of the hierarchy we have constructed:

```
Level 1 - Danny our DBA
    Level 2 – Pat
            Level 3 – Morgan
    Level 2 – Chris
            Level 3 – Kelly
    Level 2 – Van
            Level 3 – Sam
```

Van is responsible for **Vendor** data. Sam must be able to INSERT, DELETE, and UPDATE to help with the maintenance of the **Vendor** table. Van executes this command:

GRANT INSERT, UPDATE, DELETE ON VENDOR to Sam3.

6.4. USING GRANT ALL

When dealing with object privileges, there is another shortcut using the ALL keyword. We have several instances where we might want to:

GRANT SELECT, INSERT, DELETE, UPDATE on *some-table* to *someone*.

You could use the ALL keyword here and simplify the individual grants like this:

GRANT ALL on *some table* to *someone*.

The ALL keyword is part of the ANSI-92 standard including the privileges SELECT, INSERT, DELETE, UPDATE, and REFERENCES but does not include ALTER. The REFERENCES constraint allows those with the privilege to create referential constraints on the table.

ALTER is a powerful statement allowing the adding, modifying, or deleting columns from a table. ALTER may also be used to modify constraints. ALTERing a table changes the structure of the database. SELECT, INSERT, UPDATE, and DELETE may change the data in the database; however, ALTER is a system privilege and the others (INSERT, SELECT ..) are object privileges. Although ALTER can be GRANTed to others, ALTERing a table should likely be left to the creator/owner of the table. Remember the owner of a table has all privileges on the table.

Finally, there is one more object constraint we haven't mentioned yet -- EXECUTE. The EXECUTE privilege applies to functions, procedures, and packages. Since we are dealing with simple table privileges at this point, we will defer discussion of EXECUTE keyword.

Pat and Chris need to execute GRANT ALL to their persons at Level 3.

Exercises for Chapter 6

Ex 6.1.
Connect as Van, Chris, and Pat and execute appropriate table backups for **Vendor**, **Customer**, and **Product**, respectively.

Ex 6.2.
Connect as Van, Chris, and Pat. Create and execute create audit scripts for **Vendor**, **Customer**, and **Product**, respectively. Run the scripts for each Level 2 person. Level 2 persons should present their audit script to the other users and be open to suggestions for improvement.

Ex 6.3.

Connect as Van, Chris, and Pat and execute appropriate GRANTs to Sam, Kelly, and Morgan.

Ex 6.4.

Each Level 3 person should perform several DML commands. In doing these commands, they should record what they did and when they did it. For example, have Kelly INSERT a customer, Sam UPDATE a vendor row, and Morgan DELETE a product. Then, have Sam DELETE two vendors, etc.

Ex 6.5.

REVOKE the UPDATE privilege from Pat. Then, see if the UPDATE has cascaded to Morgan.

(1) Connect as Morgan and have Morgan try to UPDATE a row in **Product**.
(2) Display the **Dictionary** table ALL_TAB_PRIVS by writing this script:

```
/* Query_ALL_OBJ_PRIVS.SQL */
/* June 21 2021 */
COLUMN GRANTEE                            FORMAT A8
COLUMN OWNER                              FORMAT A8
COLUMN TABLE_NAME
    FORMAT A9
COLUMN GRANTOR                            FORMAT A8
COLUMN PRIVILEGE    HEADING "PRIV"        FORMAT A9
COLUMN GRANTABLE    HEADING "G-able"      FORMAT A6
COLUMN Hierarchy    HEADING "Hier"        FORMAT A4
SELECT * FROM       ALL_TAB_PRIVS;
```

SQL> @ Query_ALL_OBJ_PRIVS

Ex 6.6.

Each user should have the Level 3 persons show what they did to the main table. Then, the Level 2 user should run the audit script again after the exercises above are completed. Do the counts and amounts balance in each main area? If not, why not?

Ex 6.7.

Restore the three main tables to the values they had before these exercises. All Level 2 users should have created backup tables so re-establishment of the tables prior to this exercise should be simply reversing the backup procedure:

DELETE FROM Vendor;
INSERT INTO Vendor SELECT * FROM Vendor_bak0807;

This assumes the backup prior to this exercise was *table*_bak0807.

Chapter 7

EXPANDING THE DATABASE

We now have our three main tables: **Vendor**, **Product**, and **Customer**. We have also established our users and have set up privileges for each user to view or manipulate tables. Per exercises at the end of Chapter 6, Level 2 users and Danny have audit scripts on the main tables. Each Level 2 user has versioned backups of the main tables. In this chapter, we will assume everything is in working order and see if any problems arise by adding two tables and querying the database.

When a person builds a piece of electronic equipment, the moment arrives when power is applied -- electronic circuit builders call this the "smoke-test." If the circuit smokes, something is clearly wrong. In this chapter, we apply the smoke-test to our little database by creating linking tables to complete the M:N relationships between **Customer** and **Product** as well as between **Vendor** and **Product**.

Many customers buy many products. This infers an M:N relationship between customers and products. Normally, an M:N relationship such as **Customer:Product** is realized using a linking table containing the key of the **Customer** table, the key of the **Product** table, and some "intersection data." Here, the intersection data would be at least the price paid for the product; it could contain more data.

7.1. GETTING STARTED WITH LINKING TABLES

To refresh your memory, here are the tables we have created:

```
SQL> DESC Vendor
Name                                     Null?    Type
---------------------------------------- -------- ---------------------------
Vendor_ID                                         NUMBER(3)
Vname                                             VARCHAR2(20)
Vaddress                                          VARCHAR2(30)
Vcity                                             VARCHAR2(20)
Vstate                                            CHAR(2)
Vzip                                              CHAR(5)
Vphone                                            CHAR(10)

SQL> DESC Product
Name                                     Null?    Type
---------------------------------------- -------- ---------------------------
Product_ID                                        NUMBER(4)
Pname                                             VARCHAR2(20)
Ptype                                             VARCHAR2(20)
QOH                                               NUMBER(38)
Price                                             NUMBER(7,2)
Item_type                                         VARCHAR2(17)

SQL> DESC Customer
Name                                     Null?    Type
---------------------------------------- -------- ---------------------------
Customer_ID                                       NUMBER(3)
Cname                                             VARCHAR2(20)
Caddress                                          VARCHAR2(30)
Ccity                                             VARCHAR2(20)
Cstate                                            CHAR(2)
Czip                                              CHAR(5)
Cphone                                            CHAR(10)
```

We have created the three main tables and set them up so everybody in the group can SELECT from all of these tables. Furthermore, we have created synonyms for all tables in all accounts. The next step is to create some intersection tables to link these main tables together.

We assume M:N relationships for **Customer:Product** and **Product:Vendor** because, *Many customers buy Many products and Many products are bought by Many customers. Many products are purchased from Many vendors and Many vendors sell Many products to Hardware City.* The task is to set up the linking or intersection tables for these two intersection relationships and to populate them. These intersection tables should be created at Level 2, and the question must be asked "Who will be responsible for these two tables?" And, "Why Level 2?"

At this point, a management decision must be made. The DBA makes the choice. Level 2 should manage intersection data because Level 2 users control **Customer**, **Product,** and **Vendor**. It, therefore, seems appropriate for the programmers at Level 2 to handle the intersection tables.

To bridge customers and products, the most appropriate people to deal with the intersection data would be Chris or Pat. Danny appoints Pat to be the caretaker of the **Buy** table.

Pat then thinks, "What is required for the link between customers and products (the **Buy** table)?" The linking table will have a concatenated key (*Customer_ID, Product_ID*) with foreign key integrity constraints such that all rows in **Buy** have customers and products already in the database. Further, intersection data will be quantity bought, price paid per item, and a date. The description of **Buy** in Pat's view would look like this:

Buy (Cust_ID, Prod_ID, Qty, Price, Dte)

We will illustrate the basic version of table creation without constraints and leave the complete and proper creation of **Buy** as an exercise. "Complete and proper" means all constraints are placed on the table, and each constraint is properly named and tested. We shortened the names of the attributes in **Buy** for *Customer-id* to *Cust_ID* and *Product-id* to *Prod_ID*, so these attributes would have a different name from the referenced primary keys.

Also, what does Pat need to do to bring Chris into the process and what other privileges need to be GRANTed?

Here is a non-constrained version of the creation of **Buy**:

```
SQL> CREATE TABLE Buy(
Cust_id         NUMBER(3),
Prod_id         NUMBER(4),
Qty             NUMBER(6),        -- Quantity purchased
Price           NUMBER(7,2),      -- Price paid by Customer
Dte             DATE              -- Date of transaction
);

Table created.
```

Now suppose we have the following transactions:

```
INSERT INTO Buy VALUES(340,5000,1,6.35, TO_DATE('06/21/2021','mm/dd/yyyy'))
INSERT INTO Buy VALUES(335,2000,3,4.35, TO_DATE('06/21/2021','mm/dd/yyyy'))
INSERT INTO Buy VALUES(340,3000,10,1.54, TO_DATE('06/21/2021','mm/dd/yyyy'))
INSERT INTO Buy VALUES(345,2000,5,4.05, TO_DATE('06/21/2021','mm/dd/yyyy'))
INSERT INTO Buy VALUES(335,3000,17,1.65, TO_DATE('06/21/2021','mm/dd/yyyy'))
INSERT INTO Buy VALUES(350,3000,10,1.48, TO_DATE('06/21/2021','mm/dd/yyyy'))
INSERT INTO Buy VALUES(345,4000,2,8.02, TO_DATE('06/21/2021','mm/dd/yyyy'))
INSERT INTO Buy VALUES(335,1000,1,6.35, TO_DATE('06/21/2021','mm/dd/yyyy'))
```

While the attributes Cust_ID, Prod_ID, Qty, and Price are simply numbers, the insertion of a date into the table is illustrated with a specific date format using the TO_DATE function to insure uniformity.

```
SQL> SELECT * FROM Buy;
```

Customer_ID	Product_ID	QTY	Price	DTE
-----------	----------	--------	-------	----------
340	5000	1	6.35	21-JUN-21
335	2000	3	4.35	13-JUN-21
340	3000	10	1.54	19-JUN-21

345	2000	5	4.05	16-JUN-21
335	3000	17	1.65	22-JUN-21
350	3000	10	1.48	21-JUN-21
345	4000	2	8.02	21-JUN-21
335	1000	1	6.35	21-JUN-21

8 rows selected.

The other intersection data table is **Product-Vendor** which we'll call **Supply**. For **Supply**, Danny will appoint Van as the caretaker. Here is a script to create and populate **Supply**:

```
/* This script creates an unconstrained version of the intersection table
Supply
    Adding constraints is an Exercise at Chapter's end
    This table connects Vendor and Product which is an M:N relationship
*/
    /* June 22 2021 */
    CREATE TABLE Supply (
    Product_id      NUMBER(4),
    Vendor_id       NUMBER(3),
    Qty             NUMBER(6),       -- Quantity supplied
    Price           NUMBER(7,2),     -- Price paid for item
    Dte             DATE             -- Transaction date
    );

    /* populate the Supply table Pat2 June 22, 2021 */
    INSERT INTO Supply VALUES
(2000,100,30,3.25, TO_DATE('06/03/2021','mm/dd/yyyy'));
    INSERT INTO Supply VALUES
    (2000,200,60,3.05, TO_DATE('06/05/2021','mm/dd/yyyy'));
    INSERT INTO Supply VALUES
    (1000,400,10,8.25, TO_DATE('06/01/2021','mm/dd/yyyy'));
    INSERT INTO Supply VALUES
```

(1000,100,20,8.45, TO_DATE('06/15/2021','mm/dd/yyyy'));
INSERT INTO Supply VALUES
(3000,100,200,1.25, TO_DATE('06/03/2021','mm/dd/yyyy'));
INSERT INTO Supply VALUES
(3000,200,300,1.20, TO_DATE('06/14/2021','mm/dd/yyyy'));
INSERT INTO Supply VALUES
(3000,500,1000,1.08, TO_DATE('06/20/2021','mm/dd/yyyy'));
INSERT INTO Supply VALUES
(4000,400,40,6.25, TO_DATE('06/13/2021','mm/dd/yyyy'));
INSERT INTO Supply VALUES
(4000,300,25,6.45, TO_DATE('06/10/2021','mm/dd/yyyy'));
INSERT INTO Supply VALUES
(4000,200,20,6.85, TO_DATE('06/20/2021','mm/dd/yyyy'));
INSERT INTO Supply VALUES
(5000,100,35,5.85, TO_DATE('06/08/2021','mm/dd/yyyy'));
INSERT INTO Supply VALUES
(5000,400,75,5.75, TO_DATE('06/12/2021','mm/dd/yyyy'));

SQL> SELECT * FROM Supply;

Product_ID	Vendor_ID	QTY	Price	DTE
2000	100	30	3.25	03-JUN-21
2000	200	60	3.05	05-JUN-21
1000	400	10	8.25	01-JUN-21
1000	100	20	8.45	15-JUN-21
3000	100	200	1.25	03-JUN-21
3000	200	300	1.2	14-JUN-21
3000	500	1000	1.08	20-JUN-21
4000	400	40	6.25	13-JUN-21
4000	300	25	6.45	10-JUN-21
4000	200	20	6.85	20-JUN-21
5000	100	35	5.85	08-JUN-21
5000	400	75	5.75	12-JUN-21

7.2. INTEGRITY CONSTRAINTS

We have given you the basic version of the CREATE TABLE commands for **Buy** and **Supply**. In Exercise 7.1, you will need to re-do the creation scripts to include integrity constraints. In dealing with a database, it is vital there be controls on the data inserted into the database. The transaction control methods we will enhance later (more audits) are an after-the-fact database-checking technique. The constraints included here are before-the-fact integrity enforcement.

7.3. SOME BEGINNING QUERIES

Now, let us try a few queries to test out privileges and integrity constraints. First, we will connect as Pat2, show the intersection tables, and then GRANT privileges to the other users. All users will get SELECT on the table **Buy**. In addition, Chris2 will get UPDATE, DELETE, and INSERT on **Buy** because Chris manages **Customer**. Van2 will GRANT privileges on **Supply**.

CONN Pat2/Pat

Connected to:
Oracle Database 11g Enterprise Edition Release 11.2.0.1.0 - 64bit Production
With the Partitioning, OLAP, Data Mining and Real Application Testing options

SQL> SELECT * FROM Buy;

Customer_ID	Product_ID	QTY	Price	DTE
-----------	----------	---------	--------	---------------
340	5000	1	6.35	21-JUN-21

335	2000	3	4.35	13-JUN-21
340	3000	10	1.54	19-JUN-21
345	2000	5	4.05	16-JUN-21
335	3000	17	1.65	22-JUN-21
350	3000	10	1.48	21-JUN-21
345	4000	2	8.02	21-JUN-21
335	1000	1	6.35	21-JUN-21

8 rows selected.

SQL> SELECT * FROM Supply;

Product_ID	Vendor_ID	QTY	Price	DTE
2000	100	30	3.25	03-JUN-21
2000	200	60	3.05	05-JUN-21
1000	400	10	8.25	01-JUN-21
1000	100	20	8.45	15-JUN-21
3000	100	200	1.25	03-JUN-21
3000	200	300	1.2	14-JUN-21
3000	500	1000	1.08	20-JUN-21
4000	400	40	6.25	13-JUN-21
4000	300	25	6.45	10-JUN-21
4000	200	20	6.85	20-JUN-21
5000	100	35	5.85	08-JUN-21
5000	400	75	5.75	12-JUN-21

12 rows selected.

To simplify the task of GRANTing SELECT to all users, Danny1 creates a table of users in the Hardware_City database and then inserts all usernames into it. Here is an example of a different way to create a script without the host, cat, exit scenario we used earlier.

```
/* The creation of the Buy table from the account of Pat2 */
/* GRBuy script*/
CREATE TABLE HC_USERS1    (UNAME VARCHAR2(20));
INSERT INTO HC_USERS1     VALUES ('Pat2');
INSERT INTO HC_USERS1     VALUES ('Chris2');
INSERT INTO HC_USERS1     VALUES ('Van2');
INSERT INTO HC_USERS1     VALUES ('Kelly3');
INSERT INTO HC_USERS1     VALUES ('Morgan3');
INSERT INTO HC_USERS1     VALUES ('Sam3');
INSERT INTO HC_USERS1     VALUES ('Danny1');
SPOOL GRBuy.SQL
```
SELECT 'GRANT SELECT ON Buy TO '||UNAME||';' FROM HC_USERS1;
```
SPOOL OFF

@GRBuy
```

When the GRBuy script executes, every user in the HARDWARE_CITY system will be in the table **USERS1**. Then, when the script is executed, each user is GRANTed SELECT on **Buy**.

Finally, each user will have to create a synonym for **Buy** and **Supply** to simplify the usage of the tables.

The easiest way to do this is to connect to each user, one at a time, and CREATE the SYNONYM in each account. This must be done after they have been granted SELECT on the **Buy** and **Supply** tables.

And, here is our data at this point:

SQL> @Vendor_query

Vid	Vname	Vaddr	Vcity	ST	Vzip	Vphone
100	Acme Supply	123 PTree St.	Atlanta	GA	30002	4045551325
200	Norris Pumps	6 Palafox St.	Pensacola	FL	32501	8505551313
300	Petro Pipes	99 University Av	Tuscaloosa	AL	35402	3345551966
400	Hardy Hardware	33 Tornillo Dr.	Los Angeles	CA	91007	6265551234
500	Snow Screws	82 Fargo Rd.	Shakopee	MN	55378	9525556789

SQL> @customer_query

Cid	Cname	Caddr	Ccity	ST	zip	Phone
330	Abbie Walker	1988 Druid Hwy.	Cumming	GA	30041	6785557272
335	Matt Houston	15823 Fish Lane.	Gainesville	GA	30508	7705550001
340	Daphne Jayne	1 Small Ct.	Dunwoody	GA	30347	6785552016
345	Ellie Texann	2014 Newly Blvd.	Brookhaven	GA	30319	4045553333
350	Penny Penn	77 Nopound St.	Atlanta	GA	30303	4045559996

SQL> @Product_query

PID	Pname	Ptype	QOH	Price	Item_type
1000	Saw Blades	Tools	100	9.85	PKG
2000	Paint Buckets	Paint	452	3.95	ITEM
3000	Sheet Metal Screws	Hardware	12500	1.45	PKG OF 5
4000	Wall Sockets	Electrical	124	7.68	ITEM
5000	Citronella	Chemicals	25	6.24	Candle

Now suppose Kelly wants to know what customer "Penn" bought?

```
SQL> l
 1  SELECT   c.Cname, p.Pname
 2  FROM     Buy b, Customer c, Product p
 3  WHERE    b.Customer_id = c.Customer_id -- equi join
 4  AND      b.Product_id = p.Product_id -- equi join
 5* AND      c.Cname like '%Penn%'
```

Cname	Pname
Penny Penn	Sheet Metal Screws

Suppose Morgan asks, "Which customer spent the most money?"

First, we look at the table with the information we seek:

```
SQL> SELECT   Customer_id, Qty, Price, Qty*Price Spent
  2  FROM     Buy;
```

Customer_ID	QTY	Price	SPENT
340	1	6.35	6.35
335	3	4.35	13.05
340	10	1.54	15.4
345	5	4.05	20.25
335	17	1.65	28.05
350	10	1.48	14.8
345	2	8.02	16.04
335	1	6.35	6.35

The answer is Customer_id = 335 found with this query:

```
COLUMN cname FORMAT a15
SELECT    x.Customer_id, c.Cname, x.Spent
FROM      Customer c,
(SELECT   Customer_id, Qty, Price, Qty*Price Spent
   FROM          Buy
   WHERE         Price*Qty =
   (SELECT       MAX(spent) FROM
         ( SELECT   Customer_id, Qty, Price, Qty*Price Spent
           FROM  Buy)
)) x
WHERE x.Customer_id = c.Customer_id;
```

7.4. AUDITING QUERIES AND BACKUPS – VERSION 2

Auditing is necessary for the database to be maintained as a stable and correct entity. Auditing infers either an undo or backup system is available to repair incorrect changes. We introduced audits and backups in Chapter 6. We will now approach the audits and backups more formally. First, we will look at auditing and then backups.

7.4.1. Auditing

This section is about auditing the database by performing counting and summing queries. Since we will be checking the database contents, what is done if an auditing query tells us something is awry?

In any database, changes will be made: rows will be added, deleted, and changed. As changes are made, there should exist a separate table to record the changes made, who made them, and when. In Chapter 6, we suggested the person who made changes report the changes to the Level 2 person. To be more formal about the process of modifying the main tables, there has to be an audit trail to tell Level 2 users how their main table was modified. So, in addition to the main tables for each Level 2, there has to be a change-log table to accompany the main table. Here are some examples of change-log tables for the **Customer** table:

Customer_insert   (New_Cid,   Added_by,   Added_when_day, Added_when_time)

Customer_update   (Cid,   Changed_by,   Changed_when_day, Changed_when_time, Field_changed)

Customer_delete   (Old_Cid,   Deleted_by,   Deleted_when_day, Deleted_when_time)

If at the end of day Chris runs an audit query for **Customer** and finds the count of customers is different from yesterday, the change log should reflect who added or deleted a customer and when. Here, because we are looking at a somewhat simplistic version of auditing, it would be incumbent on an updater to record the action in the appropriate table. Triggers (Chapter 8) would be far superior for auditing because no overt action on the part of the person making the change would be required. For now, we will assume someone changes the **Customer** table and dutifully records the action in the change log. If the number of customers balances, then the audit may be over for today. If the counts do not match, there should be a trail to the person who made whatever change there was. It would be the responsibility of the Level 2 user to verify the changes were correct as well as undo incorrect changes and/or use a backup version of a table to figure out what happened.

Going a little deeper, Chris could design a query to count each field value in the main **Customer** table. Suppose Chris knew there were 25 distinct zip codes at the beginning of the day and 26 distinct zip codes at the end of the day. This would prompt Chris to look at the change-log tables to learn who modified the **Customer** table.

To be even stricter, Chris could approve or disapprove changes before they were made and then would be able to see the changes were consistent with the current version of the **Customer** table. There would have to be a system in place where a petition to change was made by someone, and where Chris approved or disapproved the change and finally verified the change made was valid. Chris could then enter the change data into the change log. This stricter version would likely only be workable if the number of changes per day were few.

After checking the change logs, Chris could archive the change logs, update the backup of **Customer**, and be ready for the next day. Archiving backups also involves versioning as we discussed before where the month and day were appended to the backup table. If a problem occurred, a backup could replace the changed table, and the person who made the change would have to re-petition to change the database the next day.

As an example of an auditing query, it is important to check if a customer were deleted, there must be a count of customers before and after the deletion. Further, there must be a check that the customer is deleted from all intersection tables and the new and old sums of products sold balance. You'd have to check whether the referential integrity constraints worked as designed. Remember referential integrity constraints for DELETE may be defined as RESTRICT, SET NULL, or CASCADE.

7.4.2. Backup

A suggested backup scenario could be this:

August 7:

CREATE TABLE Customer_bak0807 as
    SELECT * FROM Customer;

On August 8, the backup table would be called **Customer_bak0808**.

How long are backups kept? It depends on DML traffic and when the Level 2 person is sure all is stable.

One last point -- backup tables are for emergency use. Absolutely no one has access to these tables other than the Level 2 person who maintains them.

Exercises for Chapter 7

Ex 7.1.
Modify and execute the scripts to create **Buy** and **Supply** starting with the above models. Add all necessary integrity constraints on every attribute in the create scripts including limits on amounts which can be inserted into the table for quantity and price. Suppose quantity must be between zero and 1000 and price between fifty cents and twenty dollars. Here is a checklist of constraints:

PRIMARY KEY
UNIQUE KEY
FOREIGN KEY (REFERENCES)
CHECK CONSTRAINT
NOT NULL

Be sure to include appropriate REFERENTIAL constraints in the scripts. More on referential constraints may be found in [7.1].

Expanding the Database

Ex 7.2.

Pat and Van: Create a script for granting SELECT to all users on **Supply** and **Buy**.

Ex 7.3.

Connect to each user and create a synonym for **Buy** and **Supply**.

Ex 7.4.

Find the name of the vendor who sells the least and the name who sells the most amount of stuff to Hardware_City. Least amount = MIN(Quantity *Price)

Ex 7.5.

Find the average amount spent by all customers on each product.

Ex 7.6.

(1) If you have not done so, backup the **Buy** table.
(2) Create a table and script for **Buy** updates (**Buy_update_log**).
(3) Write and execute an audit script for **Buy** counting the sum of QTY and Price in the **Buy** table. Call the script **Buy_audit**.
(4) Have Customer 350 buy 5 items of product 2000 for $6.64. Insert the new purchase into the **Buy** table. This is a simple INSERT command.
(5) Update the **Buy_update_log**, inserting the values from (d).
(6) Execute the audit script, **Buy_audit**.

If this is all done correctly, you should have:

1. The result from (c) before the purchase.
2. The information about the purchase entered into the **Buy_update_log**.
3. The result from re-execution of **Buy_audit**.

The difference in **Buy_audit** values should equal the values in the **Buy_update_log**.

Ex 7.7.
Repeat Exercise 7.6 with a similar transaction for the **Supply** table. This time, issue an UPDATE rather than an INSERT. You will have to create an update log.

Ex 7.8.
Repeat Exercise 7.6 with another transaction for the **Buy** table. This time, DELETE two rows in **Buy**. Be sure to create a delete log as part of the exercise.

Ex 7.9.
Create a backup of the main tables. Give no one the privilege of viewing backups. Write a script reporting the SUM, AVERAGE, MIN, and MAX value of each numeric attribute in all the tables we have in our database. Create a change log for each of the main tables as well as INSERT, UPDATE, and DELETE values in each of the main tables. When you are finished the updates, backup the change log and then audit your changes. To make this exercise more realistic, allow others to do the updates and then do your audits.

Ex 7.10.
We have done some preliminary auditing and the question "What if the database is not in a consistent state?" arises. The answer lies in being able to reconstruct the tables using a backup system. Practically, if the tables were very large, it might be more prudent to undo bad transactions. Here our tables are small, and we will use the backup approach. Whether a complete backup or an undo technique is taken to repair the database, a backup version of each table is necessary.
Connect as each Level 2 user and create a backup of **Vendor**, **Customer**, and **Product**. If this was done previously, verify via auditing the backups are correct and in good order -- they should be identical to the

current version of the main table. Each Level 2 user who is responsible for the intersection tables should also create a backup of the intersection data -- Pat for **Buy** and Van for **Supply**. Van and Pat should have these backup tables hidden. One more note on backup tables: As mentioned, these hidden backup tables should be versioned. The timing would depend on how often tables are checked and audited, but the versioning might go like this:

Year 1 backups - the original data as of Jan 1, 20xx
Month n backups - the data as of Jan, Feb, March, etc.
Week m backups - Jan week1, ending Jan nn
Day k backups - mm/dd/yyyy backup at end of day.

In this way, should (say) Van discovers a problem in the **Vendor** table, Van can go back and see where the data was consistent, lock down **Vendor** (Chapter 10), figure out which data needs to be repaired, restore the **Vendor** table to the point of validity, and unlock it.

REFERENCES

[7.1]   Burleson Consulting, "Foreign Key Constraint Tips," http://www.dba-oracle.com/t_constraint_foreign_key.htm

Chapter 8

TRIGGERS TO ENFORCE AUDITING

In the previous chapter, we advocated auditing. With multiple users being able to change information in multiple tables, we suggested a change log be introduced for each table in the database. Before it was incumbent on the person changing data to put the necessary information into the change log. With triggers, this change log is kept is automatically.

8.1. WHAT IS A TRIGGER?

A trigger is a PL/SQL procedure executing when a table-modifying event (such as an INSERT, DELETE, or UPDATE) occurs. Triggers may be fired for each **row** affected (FOR ANY ROW [WHEN] ...), or may be fired at the **statement** level for an INSERT, UPDATE, or DELETE operation. Triggers may be fired **before** or **after** an event (a triggering action) takes place. There are twelve categories of triggers (BEFORE/AFTER, row/statement, UPDATE/INSERT/DELETE)[8.1]:

BEFORE UPDATE row-level
BEFORE UPDATE statement-level
AFTER UPDATE row-level

AFTER UPDATE statement-level
BEFORE INSERT row-level
BEFORE INSERT statement-level
AFTER INSERT row-level
AFTER INSERT statement-level
BEFORE DELETE row-level
BEFORE DELETE statement-level
AFTER DELETE row-level
AFTER DELETE statement-level

With Hardware_City, we want to audit changes in our tables made by the Level 2 and 3 people. We will use AFTER triggers at the row level for INSERT, UPDATE, and DELETE operations. AFTER triggers fire after a change takes place, e.g., an UPDATE of a field in the **Customer** table (*note*: The **Customer** table has been created in an earlier chapter). A ROW level trigger operates on a change in any row rather than a global change to a table.

A BEFORE trigger fires before a change is attempted. The trigger may either deny the change entirely or be used to set specific values based on the values supplied for an INSERT or an UPDATE.

With the AFTER trigger, we are dealing with auditing; we presume an update has been executed (an INSERT, UPDATE, or a DELETE operation).

In the previous chapter, we suggested the creation of change-log tables like these for **Customer**:

Customer_insert   (New_Cid,   Added_by,   Added_when_day, Added_when_time)

Customer_update   (Cid,   Changed_by,   Changed_when_day, Changed_when_time, Field_changed)

Customer_delete   (Old_Cid,   Deleted_by,   Deleted_when_day, Deleted_when_time)

In this chapter, we will create triggers to deal with each of these change logs. First of all, before we can create a trigger, guess what? We have to be GRANTed the privilege to do so. Danny must GRANT CREATE TRIGGER to ROLE2. Since all Level 2 users have already been granted ROLE2, when Danny GRANTs this new privilege to ROLE2, all Level 2 users will inherit the privilege.

We will begin with a simple example of a trigger, and then we will embellish it. Rather than use the versions of change logs above, it will be easier to mimic the **Customer** table in the code for the trigger; we can simply put in the changed values (:new) along with "what was" (:old). Here is a better change-log table for UPDATEs including all fields in the **Customer** table:

```
CREATE TABLE          Customer_UPDATE_LOG
(Customer_ID          NUMBER(3),
CNAME                 VARCHAR(20),
CADDRESS              VARCHAR(30),
Ccity                 VARCHAR(20),
Cstate                CHAR(2),
Czip                  CHAR(5),
Cphone                CHAR(10),
CHANGED_BY            VARCHAR(20),
CHANGED_WHEN          DATE);
```

Then, the trigger for auditing UPDATEs on **Customer** could look like this:

```
CREATE OR REPLACE TRIGGER Change_log_Customer_Update
AFTER UPDATE ON Customer
FOR EACH ROW
BEGIN
INSERT INTO Customer_Update_Log VALUES
(:old.customer_id,
:old.cname,
```

:old.caddress,
:old.ccity,
:old.cstate,
:old.czip,
:old.cphone,
user,
sysdate);
END;
/

Then,

CREATE OR REPLACE TRIGGER Change_Log_Customer_Update
AFTER UPDATE ON Customer
FOR EACH ROW
BEGIN
INSERT INTO Customer_Update_Log VALUES(
 :old.customer_id, :old.cname,
 :old.caddress,:old.ccity,
 :old.cstate, :old.czip,
 :old.cphone, user, sysdate);
INSERT INTO Customer_Update_Log_Values(
 :new.customer_id, :new.cname,
 :new.caddres, :new.ccity,
 :new.cstate, :new.czip,
 :new.cphone, user, sysdate);
END;
/

To see how this works, the table called **Customer_UPDATE_LOG** is created using the aforementioned CREATE TABLE **Customer_UPDATE_LOG**.

Then, we use PL/SQL to create the trigger monitoring UPDATE changes in the **Customer** table:

CREATE OR REPLACE TRIGGER Change_log_Customer_Update
AFTER UPDATE ON Customer
... *the rest is the same as above*

When created, the trigger is enabled. When the trigger fires, it updates the log table putting the :old values into it.

```
...
INSERT INTO Customer_UPDATE_LOG
VALUES (:old.customer_id,
        :old.CNAME, ...
```

Then, the trigger inserts the :new values:

```
/* Then, put in the new values: */
INSERT INTO Customer_UPDATE_LOG
VALUES(:new.customer_id,
        :new.CNAME,
        :new.CADDRESS, ...
```

In this case, the name of the script creating this trigger was chosen to be "Ctrigu."

SQL> @Ctrigu

Trigger created.

The actual name of the trigger is Change_log_Customer_Update.
The name of the script used to create the trigger is Ctrigu.
The name of the table receiving the audit information is
    Customer_UPDATE_LOG.

Now the smoke test:

Here are values in the table before updating.

SQL> COL Caddress FORMAT a20
SQL> SELECT Customer_id, Caddress FROM Customer;

Customer_ID	CADDRESS
330	1988 Druid Hwy.
335	15823 Fish Lane.
340	**1 Small Ct.**
345	2014 Newly Blvd.
350	77 Nopound St.

We will change the address of Customer 340 from "1 Small Ct." to "2 Larger Ct."

SQL> UPDATE Customer SET Caddress = '2 Larger Ct.' WHERE Customer_id = 340;

1 row updated.

SQL> SELECT Customer_id, Cname, Caddress
2 FROM customer_update_log;

CID	CNAME	CADDRESS	CHANGED_BY	CHANGED_WHEN
340	Daphne Jayne	1 Small Ct.	Chris2	30-NOV-21
340	Daphne Jayne	2 Larger Ct.	Chris2	30-NOV-21

The result of this query could be formatted differently. For example, the date attribute could be formatted to give the exact time of day and the widths of the output fields could be made larger or smaller.

What is not shown is the operation of the trigger which automatically monitored this UPDATE and placed information about the change in the **Customer_UPDATE_LOG** table.

We have seen an example of the process of trigger creation and use in this section. Our next illustration will be to embellish this trigger and then show the result.

8.2. AN EMBELLISHED TRIGGER FOR HARDWARE_CITY

While the above trigger works well, it falls a bit short of what we really want to accomplish with auditing. If we adopted the approach from above, we would have to write triggers for each table for each change. Rather, we can create a more robust trigger for each main table like this (Modeled after "A Fresh Look at Auditing Row Changes," by Connor McDonald [8.2]):

We start with a new audit table which we will use for all changes in the **Customer** table:

```
CREATE TABLE      Customer_AUDIT
(Customer_ID      NUMBER(3),
CNAME             VARCHAR(20),
CADDRESS          VARCHAR(30),
Ccity             VARCHAR(20),
Cstate            CHAR(2),
Czip              CHAR(5),
Cphone            CHAR(10),
CHANGED_BY        VARCHAR(20),
CHANGED_WHEN      DATE,
OLD_OR_NEW        CHAR(1),
TYPE_OF_CHANGE    CHAR(1));
```

In this table, we included a column for :new **or** :old and include the type of change (U = Update, I = Insert, D = Delete).

Now the trigger:

```
CREATE OR REPLACE TRIGGER Customer_audit_trigger

/* Name of trigger:  Customer_audit_trigger
   Name of audit table:        Customer_audit
   Name of the script to create the trigger:
            Ccat (short for Create_Customer_audit_trigger)
*/
AFTER INSERT or UPDATE or DELETE on Customer
FOR EACH ROW
DECLARE
   op CHAR(1) :=
            CASE WHEN updating THEN 'U'
            WHEN deleting THEN 'D'
            ELSE 'I'
      END;
BEGIN
/* Record old values in the audit table */
INSERT INTO Customer_audit VALUES
(:old.Customer_id,
    :old.CNAME,
    :old.CADDRESS,
    :old.Ccity,
    :old.Cstate,
     :old.CZip,
    :old.Cphome,
    User,
    Sysdate,'O',op); /* O for old values, op is the variable from above
*/
   /* Then, INSERT new values into the audit table
   if updating or inserting */
   IF UPDATING OR INSERTING THEN
   INSERT INTO Customer_audit VALUES
   (:new.customer_id,
   :new.CNAME,
   :new.CADDRESS,
   :new.Ccity,
   :new.Cstate,
   :new.Czip,
   :new.CPHONE,
```

Triggers to Enforce Auditing 113

user, sysdate,'N',op); /*N = New values */
END IF;
END;
/

8.2.1. Testing the Trigger

To test the trigger, we first begin by backing up the **Customer** table one more time. We will restore the original table when we are finished. Here is the result of test:

```
SQL> l
  1  CREATE TABLE        Customer_audit
  2  (Customer_ID        NUMBER(3),
  3   Cname              VARCHAR(20),
  4   Caddress           VARCHAR(30),
  5   Ccity              VARCHAR(20),
  6   Cstate             CHAR(2),
  7   Czip               CHAR(5),
  8   Cphone             CHAR(10),
  9   CHANGED_BY         VARCHAR(20),
 11   CHANGED_WHEN       DATE,
 12   OLD_OR_NEW         CHAR(1),
 13*  TYPE_OF_CHANGE     CHAR(1));

SQL> @Create_customer_audit

Table created.

SQL> @Ccat
Trigger created.

SQL> CREATE TABLE Customerbak as SELECT * FROM Customer;
```

Table created.

Using the previously stored script for viewing the contents of the **Customer** table:

SQL> @Customer_query

```
Cid   Cname           Caddr              Ccity         ST    zip     Phone
----  --------------  -----------------  ------------  --    ------  ----------
330   Abbie Walker    1988 Druid Hwy.    Cumming       GA    30041   6785557272
335   Matt Houston    15823 Fish Lane.   Gainesville   GA    30508   7705550001
340   Daphne Jayne    1 Small Ct.        Dunwoody      GA    30347   6785552016
345   Ellie Texann    2014 Newly Blvd.   Brookhaven    GA    30319   4045553333
350   Penny Penn      77 Nopound St.     Atlanta       GA    30303   4045559996
```

We now insert a new customer:

SQL> INSERT INTO Customer VALUES (331,'Mary E. Smith','11 Lala Lane', 'Los Angeles','CA','90022','4241234567');

1 row created.

Then we delete an old customer:

SQL> DELETE FROM Customer WHERE Customer_id = 340

Then, an UPDATE:

SQL> UPDATE Customer SET Czip = 33333 WHERE Czip = 30303;

1 row updated.

And now we look at the audit table:

SQL> SELECT * FROM Customer_audit;

SQL> SET LINESIZE 100

```
Cid  Cname           Caddr             Ccity          ST   Czip    CPhone        CHANGED_BY
---- --------------  ----------------  -------------  --   ------  -----         ------------
331  Mary E. Smith   11 Lala Lane      Los Angeles    CA   90022   4242345672    Chris2 NI
340  Daphne Jayne    1 Small Ct.       Dunwoody       GA   30346   4045520162    Chris2 OD
350  Penny Penn      77 Nopound St.    Atlanta        GA   33333   4045559996    Chris2 NU
```

In the **Customer_audit** table, we observe the inserted customer 331 per the line in the audit table:

331 Mary E. Smith 11 Lala Lane Los Angeles CA 90022 4242345672 Chris2 NI

NI = new data, I = INSERT

The deleted customer 340 may be noted with no new values (just the old ones):

340 Daphne Jayne 1 Small Ct. Dunwoody GA 30346 4045520162 Chris2 OD

O = old data, D = delete operation

The updated customer 350 is audited as:

350 Penny Penn   77 Nopound St.   Atlanta   GA 30303 4045559996 Chris2 OU
350 Penny Penn   77 Nopound St.   Atlanta   GA 33333 4045559996 Chris2 NU

O = Old, N = New, U = Update

There are two rows in the **Audit** table for updates, one before the UPDATE with Old values and one after the UPDATE with New values.

SQL> @customer_query

Cid	Cname	Caddr	Ccity	ST	zip	Phone
330	Abbie Walker	1988 Druid Hwy.	Cumming	GA	30041	6785557272
335	Matt Houston	15823 Fish Lane.	Gainesville	GA	30508	7705550001
345	Ellie Texann	2014 Newly Blvd.	Brookhaven	GA	30319	404 5553333
350	Penny Penn	77 Nopound St.	Atlanta	GA	33333	4045559996
331	Mary E. Smith	11 Lala Lane	Los Angeles	CA	90022	4241234567

By observation, the deleted customer 340 is now gone. Customer 331 has been INSERTed, and the Zip code 33333 replaced 30303 on customer 350, Penny Penn.

The original customer is now restored:

SQL> DELETE FROM Customer;

5 rows deleted.

SQL> INSERT INTO Customer SELECT * FROM Customerbak;

5 rows created.

SQL> @Customer_query

Cid	Cname	Caddr	Ccity	ST	zip	Phone
330	Abbie Walker	1988 Druid Hwy.	Cumming	GA	30041	6785557272
335	Matt Houston	15823 Fish Lane.	Gainesville	GA	30508	7705550001
340	Daphne Jayne	1 Small Ct.	Dunwoody	GA	30347	6785552016
345	Ellie Texann	2014 Newly Blvd.	Brookhaven	GA	30319	4045553333
350	Penny Penn	77 Nopound St.	Atlanta	GA	30303	4045559996

8.3. BEFORE TRIGGERS VS. CONSTRAINTS

The triggers we have illustrated are AFTER triggers used for auditing after-the-fact. We defined our CREATE TABLEs with CONSTRAINTs to insure integrity. Why not insure integrity with a BEFORE trigger instead of using CONSTRAINTs?

The answer is, it could be done. We will illustrate a "naked" CREATE TABLE with no CONSTRAINTs and manage the integrity with a BEFORE trigger. Here is an example:

We have a table **Sale** with an attribute *Amount* which must be between zero and 10.

In the definition of a table called **Sale**, we include a CONSRAINT like this:

```
CREATE TABLE Sale(
...
Amount    NUMBER(3) CHECK (Amount >=0 AND Amount <= 10),
Status    VARCHAR(10),
...
...
);
```

Triggers to Enforce Auditing

It is possible to ensure integrity using a trigger like this:

```
/* Name of trigger: Cintegrity */

BEFORE INSERT on Sale
FOR EACH ROW
BEGIN
IF :new.Amount >= 0 and :new.Amount <= 10 THEN
INSERT INTO Sales VALUES
(:new.attribute,
 ....
 :new.Amount
 :new.Status = 'OK'
 ....)
ELSE
(:new.attribute,
 ....
 :new.Amount = NULL
 :new.Status = 'Denied'
 ....)
END IF;
...
/
```

An integrity-checking trigger could be used instead of a CONSTRAINT, but it is discouraged by Oracle. The reason is the use of CONSTRAINTs is far more efficient. The only time an integrity checking constraint might be used is where the business rule was so complex that an ordinary CONSTRAINT would be unworkable. As an example, Amount must be less than x% of a customer's orders for the month and between 1 and 3 if the order is for some specific part.

8.4. TRIGGER CAVEATS

Before we get carried away writing triggers for many things that can happen, there are other considerations regarding them. First of all, for large databases, there may well be a performance problem with triggers. Speed of execution of SQL is often an issue, particularly if the number of users and size of the database is large.

Next, if a person writes several triggers, it is possible that one trigger causes another to fire -- a situation known as "cascading triggers." [8.3]

We recommend triggers be used for auditing and perhaps gathering statistics on transactions. They may be used for dealing with business rules so complex that they cannot be handled in any other way. In any case, triggers may be disabled and enabled and performance checked. To disable a trigger, the SQL statement would be:

ALTER TRIGGER *trigger_name* DISABLE;

To re-enable a disabled trigger, the command would be:

ALTER TRIGGER *trigger_name* ENABLE;

Exercises for Chapter 8

Ex 8.1.
Using the model in Section 8.2, create audit triggers for each of the tables we have created so far for Hardware_City: **Vendor**, **Product**, **Buy** (the intersection of **Product** and **Customer**), and **Supply** (the intersection of **Vendor** and **Product**).

Ex 8.2.
Cause the audit triggers created in Ex 8.1 to be fired using at least one DML command on each table and display the result set of the audit-checking tables.

Ex 8.3.
Create a small table called **Permission** with the following attributes:

Permission (When Date, Table Varchar(20), Grantedby Varchar(20), Grantedto Varchar(20), Operation Char(1)

Now suppose there were a business rule like this: Before any table in the database could be updated, a permission to do so must be granted by the owner of the table. For example, suppose Sam, a Level 3 user wanted to update a quantity in the **Vendor** table. Sam would have to approach Van and ask permission to do so. Van would then enter this data into the **Permission** table.

INSERT INTO PERMISSION VALUES ('7/4/17','Vendor','Van2','Sam3','u');

Construct a trigger similar to the last one we created that is a BEFORE trigger checking the **Permission** table and allowing the update by Sam if it is approved by Van. Test the trigger by Sam updating the *quantity* attribute in **Vendor** with and without permission.

Ex 8.4.
Create a table called **Cats** (*Cat_name*). Populate the table with four rows. Create the statement level framework we described in section 8.5. Execute a delete of one row and show the before and after table. Then, execute a delete of all rows and show the before and after tables.

REFERENCES

[8.1] *Practical Guide to Using SQL in Oracle*, 3rd Edition, Earp, Richard W., Bagui, Sikha S., Taylor and Francis Publishing, 2021.
[8.2] "A Fresh Look at Auditing Row Changes," by Connor McDonald, Oracle Magazine Online, March 2016, http://www.oracle.

com/technetwork/issue-archive/2016/16-mar/o26performance-2925662.html.

[8.3] See https://docs.oracle.com/cd/B19306_01/server.102/b14220/triggers.htm in which a section, "Some Cautionary Notes about Triggers," discusses cascading triggers.

Oracle goes so far as to say: "Although triggers are useful for customizing a database, use them only when necessary. Excessive use of triggers can result in complex interdependencies, which can be difficult to maintain in a large application."

Chapter 9

MULTIPLE USERS AND TRANSACTIONS

9.1. SERIALIZED TRANSACTIONS

In this chapter, we want to explore the idea of a "transaction." A transaction in SQL is a series of statements retrieving data from and/or changing the content of the database. Transactions should take place in a serial manner as though no one else interfered with the transaction in any way. "In a serial manner" implies if multiple statements are executed, then statement 2 begins when statement 1 ends; statement 3 begins when statement 2 ends, ...

The property "serializable" infers the queries and changes made to the database are made so if the person initiating a transaction were to take a before and after snapshot of affected tables, only those changes made during transaction would be evident. DML (Data Manipulation Language) Transactions change the database with INSERT, DELETE, or UPDATE commands.

The point of serializable transactions is there must be a mechanism in place so each transaction acts like no other user existed at all. Suppose user A wrote a series of commands which (a) displayed the contents of a table, then (b) inserted some data into the table, then (c) displayed the table again. You would expect all old rows were unchanged and all new rows would be observable. Hopefully, from transaction start to finish, serial transactions show only the changes made during that transaction.

The problems addressed in this chapter involve a multi-user database. It may well be impossible to isolate every database change. Let us take an example: Pat2 connects to the database. This time we will include a command to show the time:

SQL> SELECT Current_timestamp FROM Dual;

CURRENT_TIMESTAMP

17-JAN-21 11.30.37.819000 PM -06:00

Pat2 runs the script, Product_query, which looks like this:

/* Product_query written by DANNY */
/* June 21 2021 */
SET WRAP OFF
SET LINESIZE 100
SHOW USER
SELECT Current_timestamp FROM Dual;
COLUMN product_id HEADING "Pid" FORMAT 9999
COLUMN Pname HEADING "Pname" FORMAT a20
COLUMN PType HEADING "Ptype" FORMAT a20
COLUMN QOH HEADING "Qty on hand" FORMAT 99999
COLUMN Price HEADING "Price" FORMAT 99999.99
COLUMN Itemtype HEADING "Type" FORMAT a17
SELECT * FROM Pat2.Product;

SQL> @Product_query

USER is "PAT2"
SQL>

CURRENT_TIMESTAMP

20-JUL-21 05.33.04.882000 AM -05:00

SQL> SELECT * FROM Pat2.Product;

Pid	Pname	Ptype	Qty on hand	Price	Type
1000	Saw Blades	Tools	110	9.85	PKG
2000	Paint Buckets	Paint	452	3.95	ITEM
3000	Sheet Metal Screws	Hardware	12500	1.45	PKG OF 5
4000	Wall Sockets	Electrical	124	7.68	ITEM
5000	Citronella	Chemicals	25	6.24	Candle

Now imagine Morgan logs on about the same time and also executes the same script. The result will be the same for Morgan other than the timestamp.

SQL> SHOW USER
USER is "Morgan3"

SQL> SELECT Current_timestamp FROM Dual;

CURRENT_TIMESTAMP

20-JUL-21 05.34.27.665000 AM -05:00

SQL> SELECT * FROM Pat2.Product;

Pid	Pname	Ptype	Qty on hand	Price	Type
1000	Saw Blades	Tools	110	9.85	PKG
2000	Paint Buckets				

Now suppose Pat changes the price of Citronella FROM 6.24 to 6.05.

From Pat2:

SQL> UPDATE Product SET Price = 6.05 WHERE Product_id = 5000;

1 row updated.

SQL> SELECT * FROM Product WHERE Product_id = 5000;

Pid	Pname	Ptype	Qty on hand	Price	Type
5000	Citronella	Chemicals	25	6.05	Candle

After this UPDATE FROM by Pat, Morgan executes the script again:

SQL> @Product_query
SQL> SHOW USER
USER is "Morgan3"
SQL> SELECT Current_timestamp FROM Dual;

CURRENT_TIMESTAMP

20-JUL-21 05.39.12.108000 AM -05:00

SQL> SELECT * FROM Product WHERE Product_id = 5000;

Pid	Pname	Ptype	Qty on hand	Price	Type
5000	Citronella	Chemicals	25	6.24	Candle

What just happened? Pat updated the **Product** table, but Morgan sees the value unchanged. Why? When Pat connected to the database, a transaction began. Pat then initiated an UPDATE on the **Product** table; Pat did so as part of Pat's transaction. Whatever changes Pat makes in the **Product** table will not be seen by any other user until Pat concludes the transaction. How does Pat conclude the transaction? Three ways:

(1) Pat issues the command COMMIT or ROLLBACK with no SAVEPOINT clause, or
(2) Pat issues an implied COMMIT command, or
(3) Pat concludes the session wherein the UPDATE is made as Pat disconnects from Oracle.

Transaction ending situations (1) and (3) are said to be explicit because Pat **explicitly** did something to end the transaction. Situation (2) is an **implied** transaction end.

A command with an implied COMMIT is any DDL command (Data Definition Language). A DDL statement changes the structure of the database. A CREATE, ALTER, or DROP command are examples of DDL statements.

9.2. COMMIT AND ROLLBACK

COMMIT and ROLLBACK are common statements issued from the command line. COMMIT tells the system to make permanent whatever changes are made in a DML transaction. ROLLBACK undoes whatever changes are made within a transaction. The way this works is every time a transaction begins, Oracle creates a "rollback segment" in the database. This rollback segment contains data values and code that when executed will reverse the effect of a DML command. If the command ROLLBACK is issued, the rollback segment data can be used to reinstate the original data. COMMIT clears the rollback segment.

Transactions begin when a user connects to the database or when a user does not disconnect and issues a transaction-ending command.

Examples:

Scenario 1:
User A connects, a transaction X begins.
User A disconnects from the database with no transaction ending command, transaction X ends.

Scenario 2:
User A connects, a transaction Y begins.
User A does some things, then issues a COMMIT command, transaction Y ends.

Since transaction Y ended and user A did not disconnect, transaction Z begins.

Transaction Z will be active until user A ends the transaction in one of the aforementioned ways.

The point to defining a transaction is it can be rolled back or committed. The transaction we examined earlier is not over until Pat causes it to be over. Before we saw another user, Morgan, issue a command to show the contents of the **Product** table. Watch what happens when Pat ends the transaction:

SQL> SHOW USER
USER is "Pat2"
SQL> COMMIT;

Commit complete. <-- *here Pat makes the changes permanent*

SQL> SELECT Current_timestamp FROM Dual;

CURRENT_TIMESTAMP

20-JUL-21 05.41.05.698000 AM -05:00

Now, Morgan issues the same command as before but after Pat has ended the transaction.

SQL> SHOW USER
USER is "Morgan3"
SQL> SELECT Current_timestamp FROM Dual;

CURRENT_TIMESTAMP

20-JUL-21 05.41.32.883000 AM -05:00

SQL> SELECT * FROM Product WHERE Product_id = 5000;

Multiple Users and Transactions 127

Pid	Pname	Ptype	Qty on hand	Price	Type
5000	Citronella	Chemicals	25	6.05	Candle

Because Pat committed the transaction updating the **Product** table, Morgan can see the updated price for these candles.

UPDATE is a DML command. The information about the transaction can be found in the **Data Dictionary** in the **V$TRANSACTION** table; however, this table is only available to the DBA and the DBA cannot GRANT SELECT on the table to Pat2. If Pat2 really wanted to see the contents of this **Dictionary** table, Danny would have to query it for Pat.

When a transaction begins, a transaction_id is assigned. The transaction_id and accompanying **V$TRANSACTION** data does not end until we have a ROLLBACK or COMMIT (explicit or implied).

Pat initiates a transaction:

1* UPDATE Product SET Price = 6.25 WHERE Product_id = 5000
SQL> /

1 row updated.

When Pat2 COMMITs or ROLLBACKs, then the entry in **V$TRANSACTION** will go away. Pat cannot query the **V$TRANSACTION** table. If we connect as Danny, then query the tables, **V$TRANSACTION** and **V$SESSION**, the V$ tables will tell us about Pat's transaction. Here is a scenario showing **V$TRANSACTION**:

From Pat2:

SQL> SHOW USER
USER is "Pat2"
SQL> SELECT Current_timestamp FROM Dual;

CURRENT_TIMESTAMP

20-JUL-21 05.52.40.147000 AM -05:00

SQL> UPDATE Product SET Pname = 'Whatever' WHERE Product_id = 5000;

1 row updated.

SQL> SELECT * FROM Product WHERE Product_id = 5000;

Pid	Pname	Ptype	Qty on hand	Price	Type
5000	Whatever	Chemicals	25	6.05	Candle

From Danny1:

SQL> CONN Danny1/Danny

Connected.

SQL> SELECT Current_timestamp FROM Dual;

CURRENT_TIMESTAMP

20-JUL-21 05.53.10.793000 AM -05:00

SQL> SELECT
/* This script from Burleson Consulting, [9.1].
*/
t1.sid,
t1.username,
t2.xidusn,
t2.used_urec,
t2.used_ublk
FROM
v$session t1,
v$transaction t2
WHERE
t1.saddr = t2.ses_addr;

Gives:

```
    SID USERNAME  XIDUSN     USED_UREC  USED_UBLK
---------- --------- ---------- ---------- ----------
       198 PAT2          7              1          1
```

Now, Pat ROLLBACKs the UPDATE issued previously in this transaction:

From Pat2:

SQL> SHOW USER
USER is "PAT2"
SQL> ROLLBACK;

Rollback complete.

SQL> SELECT * FROM Product WHERE Product_id = 5000;

```
 Pid   Pname            Ptype                 Qty on hand   Price    Type
-----  ---------------- --------------------  -----------   -------  ---------
 5000  Citronella       Chemicals                      25      6.05  Candle
```

SQL> SELECT Current_timestamp FROM Dual;

CURRENT_TIMESTAMP

20-JUL-21 05.57.36.255000 AM -05:00

From Danny1:

CURRENT_TIMESTAMP

20-JUL-21 05.58.02.158000 AM -05:00

SQL> SELECT
t1.sid,
... *same script as above*
SQL> /
no rows selected.

After PAT2 does a ROLLBACK, the transaction is complete and no rows are selected from the **V$TRANSACTION** table.

In review, Pat's transaction, which updated the **Product** table, prevented anyone from viewing the updated row until Pat was finished, i.e., Pat's transaction was ended either explicitly or implicitly. Information about the rollback segment is kept in the **V$TRANSACTION** and **V$SESSION** tables. Ending the transaction clears the **V$** tables which contain the rollback segment information.

9.3. More on COMMIT and ROLLBACK

Since transactions may be terminated with the commands COMMIT and ROLLBACK, we need to investigate further before moving on with more transactions.

COMMIT, when issued as a command, is a strong statement by the programmer asserting the changes made are legitimate and correct.

ROLLBACK is allowable due to the way Oracle handles all transactions. When a transaction starts, an entry is made in the **V$TRANSACTION** table and any changes made by the programmer are reversible. ROLLBACK takes the database back to the place where it was before the programmer started a DML transaction.

Another example:

Pat2 is working on updating the **Product** table and means to delete the row where the *Product_id* = 3000. Supposedly, Hardware_City will no longer sell sheet metal screws. Pat2 wants to execute the command:

DELETE FROM Product WHERE Product_ID = 3000;

But instead of this, Pat2 writes the statement, gets interrupted and instead of finishing it, puts in a semicolon before the WHERE and executes this statement:

DELETE FROM Product;

Whoops! By not specifying a WHERE clause, the content of the entire table was deleted. Hopefully, there was a backup of the table; but a backup is not needed if Pat does a ROLLBACK immediately. Pat2 needs to undo the DELETE FROM Product command. Hence, Pat2 terminates the transaction with ROLLBACK and the database is restored to its original state:

SQL> DELETE FROM Product;

5 rows deleted.

SQL> @Product_query

no rows selected.

SQL> ROLLBACK;

Rollback complete.

SQL> @Product_query

Pid	Pname	Ptype	Qty on hand	Price	Type
1000	Saw Blades	Tools	100	9.85	PKG
2000	Paint Buckets	Paint	452	3.95	ITEM
3000	Sheet Metal Screws	Hardware	12500	1.45	PKG OF 5
4000	Wall Sockets	Electrical	124	68	ITEM
5000	Citronella	Chemicals	25	6.24	Candle

9.4. SAVEPOINTs AND ROLLBACK TO

While ROLLBACK is a convenient way to undo an entire transaction, we can add SAVEPOINTs to a transaction and ROLLBACK part of the transaction rather than the whole thing. The **Vendor** table represents what a vendor supplies at some asking price. Our **Supply** table indicates what we bought from each vendor and what price we actually paid for each item.

Suppose Pat was updating tables and wanted to make some changes. Here are the steps Pat follows:

(1) We note for Saw Blades/Tools, we have two vendors and currently have 100 Saw Blades on hand.
(2) The vendor named Hardy Hardware has agreed to sell us 10 more packages of Saw Blades for a price of 8.35 per package.
(3) We sell 5 packages of Saw Blades to Penny Penn.

Pat needs to update the database. In a complicated series of DML statements, it is a good idea to have a transaction plan to check the DML statements did what was intended. (*What was a DML statement again?*) Pat's transaction plan is amended like this:

1. We note for Saw Blades/Tools, we have two vendors and currently have 100 Saw Blades on hand.
    1a. Pat connects to the database and a transaction begins.
    1b. Pat tells Van to put Hardy Hardware's info into the **Vendor** table.
    1c. Pat checks the QOH (quantity on hand) from the **Product** table for Saw Blades.
    1d. The **Supply** table is checked to see how many entries we have for Hardy Hardware.
2. An entry in the **Supply** table is inserted to reflect the purchase of Saw Blades from Hardy Hardware at a price of 8.35.
3. The **Product** table needs to be updated to show we now have a sum of QOH+10 Saw Blades.
    3a. The **Product** table is queried to see it now has QOH+10 total Saw Blades.
    3b. The **Supply** table is queried to show we have one more entry from Hardy Hardware.
4 The **Buy** table is queried to find out how many entries we have for Penny Penn.
5. An entry in the **Buy** table is made reflecting the purchase by Penny Penn of 5 Saw Blades.
6. The **Product** table needs to be updated again to show that we now have QOH-5 Saw Blades.

7. The **Product** table should have a total of QOH+5 Saw Blades at this point.

8. The **Buy** table should show one more entry for Penny Penn.

9. *The transaction nears conclusion.* It is incumbent upon Pat and Van to check the results of this updating. If there is a problem in checking and balancing results, a ROLLBACK is executed. If everything balances and quantities and entries are correct, a COMMIT is executed.

10. With either COMMIT or ROLLBACK, *the transaction ends.*

Realizing this transaction involves possible pitfalls, Pat can modify the transaction plan. If there is a problem, some intermediate ROLLBACK will not destroy the whole plan.

Pat decides to include some SAVEPOINTs in the plan. The plan as outlined above now includes new actions: (Assume Hardy Hardware is in the **Vendor** table.) Here is the complete plan:

1. We note for Saw Blades/Tools, we have two vendors and currently have 100 Saw Blades on hand.

    1a. Pat connects to the database and a transaction begins.

    1b. Pat tells Van to put Hardy Hardware's info into the **Vendor** table.

    1c. Pat checks the QOH (quantity on hand) from the **Product** table for Saw Blades.

    1d. The **Supply** table is checked to see how many entries we have for Hardy Hardware.

2. The vendor named Hardy Hardware has agreed to sell us 10 more packages of Saw Blades for a price of 8.35 per package.

    2a. An entry in the **Supply** table is inserted to reflect the purchase of Saw Blades from Hardy Hardware. *The DML part of the transaction begins here.*

    2b. A SAVEPOINT, Save1, is created.

3. We sell 5 packages of Saw Blades to Penny Penn.

    3a. The **Product** table is queried to see it now has QOH+10 total Saw Blades.

    3b. The **Supply** table is queried to show we have one more entry

from Hardy Hardware.
3c. A SAVEPOINT, Save2, is added to the transaction.
3d. The **Supply** table is queried to show we have one more entry from Hardy Hardware.
3e. A SAVEPOINT, Save3, is added.

4. The **Buy** table is queried to find out how many entries we have for Penny Penn.

5. An entry in the **Buy** table is made reflecting the purchase by Penny Penn of 5 Saw Blades for 6.30.
5a. A SAVEPOINT, Save4, is added.

6. The **Product** table needs to be updated again to show we now have QOH-5 Saw Blades.
6a. The **Product** table needs to be updated again to show we now have QOH-5 Saw Blades.
6b. SAVEPOINT Save 5.

7. The **Product** table should have a total of QOH+5 Saw Blades. A value of 115 is expected.

8. The **Buy** table should show one more entry for Penny Penn

9. If all checks are correct, a COMMIT is executed. If checks fail, the transaction is ROLLBACKed. Either way, *the transaction concludes.*

If there were a problem in intermediate checking, a ROLLBACK may be executed to a specific point in the transaction to erase part of the transaction and save the prior changes. A ROLLBACK to a SAVEPOINT does not terminate the transaction.

We will follow the transaction along using SQL with annotations:

1a. SQL> SELECT QOH "Qty on hand"
FROM Product
WHERE Pname like 'Saw B%';

Gives:

Qty on hand

110

SQL> DESC Vendor

Name	Null?	Type
Vendor_ID		NUMBER(3)
Vname		VARCHAR2(20)
Vaddress		VARCHAR2(30)
Vcity		VARCHAR2(20)
Vstate		CHAR(2)
Vzip		CHAR(5)
Vphone		CHAR (10)

1b.
```
  1  SELECT      v.Vendor_id, COUNT(*)
  2  FROM        Supply s, Vendor v
  3  WHERE       s.Vendor_id = v.Vendor_id
  4  AND         v.Vname LIKE 'Hardy%'
  5* GROUP BY    v.Vendor_id
SQL> /
```

Vendor_ID	COUNT(*)
400	3

The transaction begins:

```
 1  INSERT INTO Supply VALUES
 2* (1000,400, 10,8.35,TO_DATE('06/27/2021','MM/DD/YYYY'));
```

1 row created.

SQL> SELECT * FROM Supply;

Product_ID	Vendor_ID	QTY	Price	DTE
2000	100	30	3.25	03-JUN-21
2000	200	60	3.05	05-JUN-21
1000	**400**	**10**	**8.25**	**01-JUN-21** <----
1000	100	20	8.45	15-JUN-21
3000	100	200	1.25	03-JUN-21
3000	200	300	1.2	14-JUN-21
3000	500	1000	1.08	20-JUN-21
4000	400	40	6.25	13-JUN-21
4000	300	25	6.45	10-JUN-21
4000	200	20	6.85	20-JUN-21
5000	100	35	5.85	08-JUN-21
5000	400	75	5.75	12-JUN-21

12 rows selected.

SQL> SELECT QOH "Qty on hand"
 FROM Product
 WHERE Pname LIKE 'Saw B%';

Gives:

QOH

110

SQL> SELECT   v.Vendor_id, COUNT(*)
FROM   Supply s, Vendor v
WHERE   s.Vendor_id = v.Vendor_id
AND   v.Vname like 'Hardy%'
GROUP BY   v.Vendor_id ;

Gives:

```
Vendor_ID  COUNT(*)
----------  ----------
400         3
```

SQL> INSERT INTO Supply VALUES
(1000,400, 10,8.35,TO_DATE('06/27/2021','MM/DD/YYYY'));

1 row created.

SQL> SAVEPOINT Save1;

Savepoint created.

SQL> UPDATE Product SET QOH = QOH+10;

5 rows updated.

Whoa! Pat updates the whole **Product** table rather than just the entry for Saw Blades. Pat forgot to include a WHERE clause in the UPDATE command. What to do?

SQL> ROLLBACK TO Save1;

Rollback complete.

Now Pat does the UPDATE correctly:
SQL> UPDATE Product SET QOH = QOH+10 WHERE Product_ID = 1000;

1 row updated.

SQL> SELECT QOH "Qty on hand" ... *same as before*

Gives:

QOH

120

SQL> SAVEPOINT Save2;

Savepoint created.

SQL> SELECT   v.Vendor_id, count(*)
FROM   Supply s, Vendor v
WHERE   s.Vendor_id = v.Vendor_id
AND   v.Vname LIKE 'Hardy%'
GROUP BY   v.Vendor_id;

Gives:

Vendor_ID	COUNT(*)
400	4

SQL> SAVEPOINT Save3;

Savepoint created.

SQL> SELECT   c.Customer_id, COUNT(*)
FROM   Buy b, Customer c
WHERE   b.Customer_id = c.Customer_id
AND   c.Cname LIKE 'Penn%'
GROUP BY   c.Customer_id;

Gives:

Customer_ID	COUNT(*)

350   1

So Penny Penn has only 1 purchase.

```
SQL> DESC Buy
Name                                 Null?    Type
------------------------------------ -------- ---------------------------
Customer_ID                                   NUMBER(3)
Product_ID                                    NUMBER(4)
Qty                                           NUMBER(6)
Price                                         NUMBER(7,2)
Dte                                           DATE
```

SQL> SELECT * FROM Buy;

Customer_ID	Product_ID	QTY	Price	DTE
340	5000	1	6.35	21-JUN-21
335	2000	3	4.35	13-JUN-21
340	3000	10	1.54	19-JUN-21
345	2000	5	4.05	16-JUN-21
335	3000	17	1.65	22-JUN-21
350	3000	10	1.48	21-JUN-21
345	4000	2	8.02	21-JUN-21
335	1000	1	6.35	21-JUN-21

8 rows selected.

SQL> INSERT INTO Buy VALUES (350, 1000, 5, 6.30, TO_DATE('06/27/2021','MM/DD/YYYY'));

1 row created.

SQL> SAVEPOINT Save4;

Savepoint created.

SQL> UPDATE Product SET QOH = QOH-5 WHERE Product_id = 1000;

1 row updated.

SQL> SAVEPOINT Save5;

Savepoint created.

SQL> COLUMN Pname FORMAT a20
SQL> SELECT Product_id, Pname, QOH FROM Product;

Product_ID	Pname	QOH
1000	Saw Blades	115
2000	Paint Buckets	452
3000	Sheet Metal Screws	12500
4000	Wall Sockets	124
5000	Citronella	25

```
SQL> SELECT    c.Customer_id, COUNT(*)
FROM           Buy b, Customer c
WHERE          b.Customer_id = c.Customer_id
AND            c.Cname LIKE 'Penn%'
GROUP BY       c.Customer_id;
```

Gives:

Customer_ID	COUNT(*)
350	2

SQL> ROLLBACK;

Rollback complete.

Exercises for Chapter 9

Ex 9.1.
1. To simplify this exercise, create two scripts:

a) Time_now.sql, which consists of:

SHOW USER
SELECT Current_timestamp FROM Dual;

b) CheckVtables.sql:

SELECT
t1.sid,
t1.username,
t2.xidusn,
t2.used_urec,
t2.used_ublk
FROM
v$session t1,
v$transaction t2
WHERE
t1.saddr = t2.ses_addr;

When you want to know who a user is and what time it is, execute Time_now (@Time_now). When you want to see what's in **V$TRANSACTION**, you need to connect as Danny1 and execute **CheckVtables** (@CheckVtables)

Before doing exercises on the main tables in Hardware_City, it is imperative to back up all tables. Each person responsible for each table needs to be asked whether a backup exists for each table. While the idea of rolling back a transaction seems foolproof, even foolproof situations in database tend to somehow find a way to be fouled.

2. Connect as Chris2, then execute the Time_now script.

3. Try to execute **checkVtables** from Chris2. What happens?

4. Connect as Danny1, execute Time_now, then execute **CheckVtables**.

Connecting as Danny1 may be done by logging onto UNIX and connecting as Danny1. Do not connect from the same session as when you are logged on as Chris2 because you will be disconnecting from Chris2 which is a COMMIT.

Exactly when does Chris2's transaction begin?

Currently from Danny1, do you see Chris2's transaction? If you don't see it, then Chris2 has not begun a transaction.

5. In Chris2, display the **Customer** table -- you may need to add column formatting.

6. As Chris2, UPDATE the **Customer** table by changing Daphne Jayne's name to Freddy Foster.

7. Verify your UPDATE took place by viewing the **Customer** table FROM Chris2.

8. As you are connected to Danny somewhere else, run **CheckVtables** to see if Chris2 has begun a transaction.

9. ROLLBACK your name change in the **Customer** table.

10. Verify your ROLLBACK succeeded.

11. From Danny1, verify Chris's transaction ended.

Ex 9.2.

Assume you are Van. Create a transaction plan to add a vendor to the database. Then add an entry to the **Product** table showing your new vendor supplied something. Have your transaction plan approved by your instructor. If your instructor approves the plan, connect to Oracle as Danny1, and at another station log on as Van. As you execute each step of your transaction plan, run the scripts we presented above Danny1: @Time_now and @CheckVtables, Van2: @Time_now.

Examine the results and either COMMIT or ROLLBACK as appropriate.

Ex 9.3.
Run the scripts from each account.

REFERENCES

[9.1] Burleson Consulting: http://www.dba-oracle.com/t_v_transaction.htm

Chapter 10

Concurrency

With multiple users of a database writing transactions, the target would be for each transaction to act like (or be) a serialized transaction. We have said, "The point of serializable transactions is there must be a mechanism in place so each transaction acts like no other user existed at all" (Chapter 8). Serialized transactions are not always realizable. Insisting on a transaction being serial may cause such overhead as to bog down the database. Furthermore, if a transaction is guaranteed as serialized, then other transactions may have to wait and slow the whole system. Rather than insist on serializability, we will look at ways to make transactions look and act as if they are serialized.

Delving a little deeper into the transaction itself, we will define some common terms describing transactions.

10.1. Consistency, Concurrency, and ACID Properties

We will discuss "consistency" as the same as "serializable" -- a consistent transaction will show only those changes made by the transaction as though no other transactions affected the data.

A contemporary way to look at transactions is to use what are called the "ACID properties." ACID is an acronym for Atomicity, Consistency, Isolation, and Durability. [10.1] When dealing with multi-user databases, we need to preserve the ACID properties of transactions to insure the transactions are as close to serializable as possible.

Concurrency becomes important when more than one user can access the database at about the same time. In a multi-user environment, it is unrealistic to assume every transaction will take place as a serialized transaction unless something is done to prevent interference. [10.2]

Atomicity is the property of a transaction demanding an "all or nothing" result. If we start to update tables, we expect all transactions are carried out as though nothing else were going on. In a multi-user database, this is not a given because, to insure atomicity, tables must be protected against another person changing data mid-transaction.

Consistency within the ACID view means whatever transactions are performed, the database progresses from one valid state to the next. In the previous chapter, we provided an example with SAVEPOINTs where Pat performed some DML statements. What we did was to have Pat explicitly check the tables involved for correctness. If we added ten rows to the **Product** table, did those added rows show up with all other rows remaining the same? If we sold something, was the **Product** table quantity decremented with all other quantities unchanged? If the tables **Product**, **Supply,** and **Buy** didn't provide a consistent result, then Pat2 could ROLLBACK all transactions or part of them to provide consistency. Rather than have a user like Pat explicitly deal with checking each transaction, we want to have ways to prevent other transactions from interfering with Pat's transaction before it is complete.

Isolation means if more than one transaction accesses the database, the transactions take place in such a way that one would think each transaction took place independently (in isolation). Imagine a very large database with many programmers all updating some tables at various times. For example, a transaction on the **Buy** table might affect both the **Customer** and **Product** tables. Table LOCKing is a mechanism preventing another transaction from changing a table or row while it is locked. LOCKing can be enacted in several ways; Oracle automatically does some locking when DML commands are issued.

In the real database world, it likely would be impractical for a "Pat" to simply lock everyone else out of the database until Pat's updates are completed. Hence, the subject of concurrency deals with what rules the concurrent transactions must follow. Obviously, a rule could be: When someone does DML, then all tables required by that person are locked down. This insures no one can read the tables or do anything else to them until the DML is done. This is too severe for practical purposes so some other set of rules for concurrency must be used. Oracle has systems taking care of many of the common problems which might be associated with violating ACID properties.

Durability is the property declaring once a COMMIT command is executed, the change in the database is permanent. Even if there is a system failure of some kind, the committed transaction stands.

When one person starts a transaction, an entry is made in the **V$TRANSACTION** table; other users cannot see what changes are made until the person making the changes has ended the transaction. Oracle handles most of the work on providing a consistent transaction. This is almost, but not quite, the end of this story.

A transaction for user A begins after the user connects to the database. In a multi-user environment, each database programmer is expected to end each transaction. How are transactions ended? Three ways: COMMIT, ROLLBACK, or disconnect from the database. If user A does not execute COMMIT or ROLLBACK or log off, then A's transaction may prevent anyone other than user A from seeing the result of the A's work. Furthermore, user B may not be able to perform DML actions on the table user A is changing.

10.2. Transactions Revisited

Here is an example of how we will illustrate transactions:

Illustration 1

Time	Action
00:00	Pat connects to Oracle, starts a session, and starts a transaction.
00:01	Pat updates the **Product** table -- the **Product** table is locked by Oracle (sort of.. stay tuned) before the update.
00:02	Pat disconnects from Oracle -- Pat's transaction ends, Pat's session ends.

Illustration 2

00:05	Pat connects to Oracle, starts a session, and starts a transaction.
00:06	Pat updates a row in the **Product** table -- the row in the **Product** table is locked before the update takes place.
00:07	Pat executes the COMMIT command -- Pat's transaction ends.

Since Pat didn't disconnect from Oracle, Pat's session is still in force; but, a new transaction begins for Pat due to the COMMIT.

Had Pat executed a ROLLBACK command instead of COMMIT, the result of the above would be the same; but the **Product** table would be unchanged.

There are several undesirable things possible if there were no locking (no concurrency mechanism) in place. A way of handling concurrency in Oracle is to discuss what are called "isolation levels" (as defined in the database standard for SQL92). Isolation levels in Oracle control what gets locked and for how long. There are four isolation levels involving a database transaction [10.5]:

Read Committed
Serializable
Read Only
Read uncommitted (not supported by Oracle).

Concurrency 149

From Burleson's web site: "The read committed transaction isolation level is the Oracle default. With this setting, each query can see only data committed before the query, not the transaction, began."[10.5]

There are three types of database read errors possible:

(1) Dirty reads
(2) Non-repeatable reads (fuzzy reads)
(3) Phantom reads.

A "dirty read" is also known as an "uncommitted dependency." Here is an illustration of a dirty read:

Illustration 3

00:00	Pat connects to Oracle -- Pat starts a session and starts a transaction.
00:01	Pat updates a quantity in Product 1000 in the **Product** table.
00:02	Chris reads the quantity in Product 1000 in the **Product** table.
00:03	Pat does a ROLLBACK causing Pat's transaction to commit.

If this were possible in Oracle, the data Chris read would be the result of a "dirty read." Chris could have read uncommitted data from the **Product** table because Pat's transaction was not yet committed. The isolation level allowing this dirty read would be: isolation level **read uncommitted**, which is **not** supported in Oracle.

A non-repeatable or fuzzy read illustration:

Illustration 4

00:05	Pat connects to Oracle and starts a session.
00:06	Chris connects to Oracle and starts a session.
00:07	Chris reads the quantity of Product 1000 (read #1).
00:08	Pat updates the quantity in Product 1000 in the **Product**

	table and COMMITs.
00:09	Chris reads the quantity in Product 1000 in the **Product** table again (read #2), but this time gets a different value for quantity in Product 1000.

What is happening here is Chris started a transaction, but while Chris's transaction was in progress, Pat started a transaction, updated the **Product** table and committed. In Chris's transaction the data read in read #2 would be different from read #1. In this situation Chris read committed data both times but Pat's transaction was allowed while Chris's transaction was in progress. If the isolation level read committed was in effect, this anomaly would be possible.

We'll come back to the fuzzy read in a moment; but first we will look at the other read error and then discuss what to do about preventing these errors.

An example of a phantom read:

Illustration 5

00:15	Pat connects to Oracle and starts a session.
00:16	Chris connects to Oracle and starts a session.
00:07	Chris issues a command summing the quantities of all Products. (query #1).
00:08	Pat inserts Product 2001 into the **Product** table and COMMITs.
00:09	Chris re-executes the "sum the quantities" query and gets a different result (query #1a).

In this case, Chris's query identified a set of rows to get a result. Then, Pat's insert added a row. When Chris re-executes the "sum the quantities" query, a different result appears. If the isolation level read committed were in effect, this anomaly would be possible.

So what does Chris do about the two possible errors in the default state of Oracle transaction handling? Recall, the isolation level read committed is the Oracle default. Therefore, to prevent a fuzzy read or phantom read, some other mechanism would be required of the user such as:

(a) Chris locking the **Product** table so it couldn't be updated by Pat until Chris was finished, or
(b) Chris issued a SET TRANSACTION command precluding a fuzzy read.

Issuing LOCK commands are very specific and tend to be overriding the Oracle transaction system. The more usual approach to controlling concurrency is to set user isolation level commands -- specifically, we will use SET TRANSACTION ... with an awareness of (a) Oracle defaults and (b) what other users may need to do.

The SET TRANSACTION can be used either at the session or the transaction level.

The syntax of the SET TRANSACTION command is shown in Figure 10.1 [10.6].

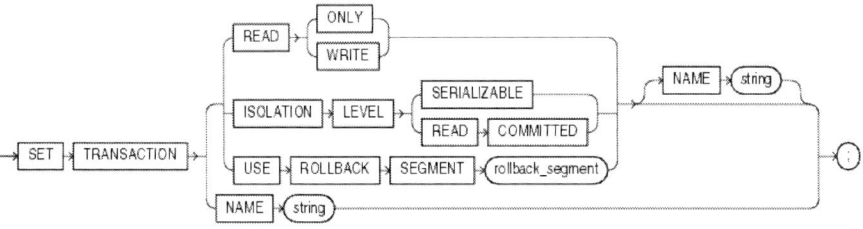

Figure 10.1. Syntax of the SET Transaction Command [10.6]

Notice there is no READ UNCOMMITTED option. READ COMMITTED, the Oracle default will prevent dirty reads. In the "case of the fuzzy read or phantom read," Chris wants to preclude these read errors, so the following options would be:

If the SET TRANSACTION command were issued at the transaction level, the isolation level persists until Chris's transaction ends. The syntax for this statement is:

SET TRANSACTION ISOLATION LEVEL *isolation level*;

Isolation level would be: READ COMMITTED, READ ONLY, or SERIALIZABLE.

READ COMMITTED is the Oracle default and would disallow dirty reads but allow fuzzy reads.

SET TRANSACTION ISOLATION LEVEL READONLY: From Burleson [10.5], "With the Read Only transaction isolation level, only data committed at the start of the transaction can be accessed by a query. No modification to the data is allowed." This level would only read what might be thought of as "old data." Any transaction outside of Chris's transaction could update the database, but Chris would see whatever was committed before the transaction started. Updates performed during Chris's transaction would not be seen. This situation produces what is called "transaction level read consistency."

SET TRANSACTION ISOLATION LEVEL SERIALIZABLE: This isolation level is the strongest of the three because it prevents dirty reads, phantom reads, and non-repeatable reads. Serializable means the transaction takes place as though no other transaction took place at all during the execution of the subject transaction.

If these isolation levels are desired for Chris's entire session, then an ALTER SESSION command would be appropriate:

ALTER SESSION SET ISOLATION_LEVEL *isolation level*;

The *isolation level* may have the same three levels described above. The only difference here is Chris's demand for SERIALIZABLE isolation might be too restrictive because no other changes to the database would be allowable while Chris's session was in force. If the isolation were set at the transaction level, once the Chris's transaction ended, the isolation level would be READ COMMITTED (the default) again. For further reading, see [10.6].

10.3. LOCKing

Locking is more particular that using isolation levels. If using isolation levels, it's almost as if you'd be working within the Oracle transaction system. Locking is more a sledge-hammer approach to controlling concurrency. Which approach is employed is very user-specific based on how many users are involved, how much isolation is required, and which tables need to be isolated. Also, there are distributed database issues beyond the introductory scope we are trying achieve.

Locking is normally discussed at levels of granularity, which tells us the level of depth at which something exists. The common levels of locking granularity in database are at the row and table levels. [10.4] If a valuable of yours is locked, you don't expect anyone to get into it. You can lock your valuables in your house at various levels of granularity -- for example, lock the house, lock a room in the house, or lock a safe within a room.

Locking in Oracle has the same sense as locking something you own. Locking is a protection mechanism. In database, locking exists at two levels of granularity -- row or table level. Normally, locking takes place at the row level. Above, we said when Pat updated the **Product** table, the data was locked by Oracle since no specific locking command was executed. If Oracle handles the situation, then the **Product** table would be locked at the row level.

If, for example, Pat updated Product 1000, then the row containing Product 1000 would be locked. If no locking command were issued, the **Product** table would not be locked; only the *row* updated in Pat's transaction would be locked.

If a user does not explicitly perform locking with LOCK commands, Oracle handles the situation and automatically locks row data when DML is performed. Locking is controllable by either a table owner or a DBA [10.2]. Explicit locking may take place at either the table or the row level. Explicit locking at the table level is provided by this statement:

LOCK TABLE *table-name* IN {SHARE | EXCLUSIVE} MODE

EXCLUSIVE mode means other users may see the contents of a table but may not perform DML on it. Some writers say an Exclusive lock will not allow anyone to see the table being locked; but in Oracle, this is not so.

SHARE mode supposedly means other users may view the contents of a table while it is locked. One might think this would mean an intermediate result might be visible, but it is not. For example, if we proceeded with the following scenario with no explicit locking, we see that Oracle takes care of locking (at the row level):

Sign on as Pat2 and Van2 on two different platforms.

In Pat2, table **Product** is updated.

Van2 queries **Product**, but Van2 does not see the Pat2 update until Pat2 ends transaction.

Van2 sees the un-updated version of **Product** (as in READONLY isolation).

Pat2 COMMITs or ROLLBACKs and Van2 can now see the table with Pat2's update in it if **Product** is re-queried by Van2.

Now suppose Pat does some explicit locking:

Pat issues this command LOCK TABLE Product in EXCLUSIVE MODE.

What Pat and Van see is the same as the above scenario.

Pat issues this command LOCK TABLE Product in SHARE MODE.

Again, Pat and Van see the same things they would see if the LOCK TABLE command were not issued at all.

Now, in the following scenario, we will see something interesting:

First, Pat2 grants Van2 the UPDATE privilege on **Product**.

Pat2 executes LOCK TABLE Product in EXCLUSIVE MODE.

Van2 issues an UPDATE command on **Product**. The command waits until Pat2 releases the lock. Then the Van2 command proceeds and the **Product** table is updated by Van.

Suppose Pat did not want Van's update to succeed. Can Pat issue a ROLLBACK to undo what Pat did? No, Pat2 would have to use the backup of **Product** to recover unless Van2 issues a ROLLBACK.

All of this so far is the same as Pat not issuing a LOCK TABLE command at all. What is different is that if no LOCK TABLE is issued, Pat can change a row and Van can change a different row. Oracle's locking system is triggered by Pat issuing an UPDATE command (like UPDATE, DELETE, or INSERT). The lock applied by Oracle is a ROW LEVEL LOCK. From the Oracle website [10.3]:

Row Locks (TX) -- A row lock, also called a TX lock, is a lock on a single row of a table. A transaction acquires a row lock for each row modified by one of the following statements: INSERT, UPDATE, DELETE, MERGE, and SELECT ... FOR UPDATE. The row lock exists until the transaction commits or rolls back.

When a transaction obtains a row lock for a row, the transaction also acquires a table lock for the table in which the row resides. The table lock prevents conflicting DDL operations to override data changes in a current transaction.

DDL operations would commands like ALTER or DROP TABLE.

If the **Product** table is explicitly locked by Pat at the table level, any change Van tries to make will wait for Dan to unlock the table.

If no *explicit* locking by Pat is in force and if Van has UPDATE privileges on **Product**, Pat can change a row and Van can change some other row because default Oracle locks are ROW locks. What if Van tries to change the same row Pat is updating? Van will wait just like when Pat issued a LOCK TABLE command.

So what does this prove? Explicit locking disallows another user from updating the table at all until the explicit lock is released. Oracle applies row level locking when someone updates a table. As we saw, Pat updates a row, Van can update another row. If Van tries to update the row Pat is working on and has not committed, Van will wait.

There are ways to disable locks on tables to enact a more complicated concurrency control method. These more complicated techniques come into play when there are many transaction conflicts to slow the system significantly. In the last chapter, we showed a transaction in slow motion, one statement at a time. In reality, a series of transaction would take place in a script or a procedure and would include triggers or the procedure itself to check for consistency problems.

10.4. DEADLOCKS AND STARVATION

A deadlock (a.k.a, "deadly embrace") is a situation in concurrency where two users hold locks such that neither user can access the other's object.

Starvation is a multi-user situation where a transaction is forced to wait due to factors outside the control of the transaction. Starvation may involve such things as buffering, disk-block control, etc. Such a situation might occur in a large database with many users and has to be resolved by restructuring the starved queries. In this small Hardware_City database, starvation is not likely. See Burleson [10.7].

For example, suppose Pat and Chris did the following:

Pat starts a transaction that updates the **Product** table where Product_id = 2000. Call this transaction 1. As we know from above, this UPDATE causes Oracle to lock the row in the **Product** table where Product_id = 2000.

Chris starts a transaction to update the **Customer** table where Customer_id = 330. This action locks the row in the **Customer** table. This is transaction 2.

Pat then issues a command to update the **Customer** table for Customer_id = 330 as part of transaction 1. (This assumes Pat has that privilege.) But, that particular row in the **Customer** table is locked by Chris's transaction 2 so Pat's transaction will wait.

Chris issues a command to update the **Product** table, Product_id = 2000 as part of transaction 2. Since Pat's transaction 1 has a lock on that row, Chris's transaction waits.

Both transactions are waiting and are deadlocked.

Oracle resolves this issue automatically by rolling back the transaction causing the deadlock. In this case, the causer of the deadlock was Pat because Pat's request to update **Customer** came after Chris's update which locked the table.

Here's how this can be followed:

First of all, two scripts were invented:

Qqc: SELECT * FROM customer WHERE Customer_id = 330;
Qqp: SELECT * FROM Product WHERE Product_id = 2000;

Ti was mentioned previously and simply displays the timestamp.

```
SQL> SHOW USER
USER is "PAT2"
SQL> @Ti
USER is "PAT2"

CURRENT_TIMESTAMP
---------------------------------------------------------------------
22-DEC-21 08.33.02.084000 PM -06:00
```

Pat connects and begins a transaction.
Here is what Pat sees for **Customer** and **Product** at 08.33.02:

```
SQL> @qqC
```

Cid	Cname	Caddr	Ccity	ST	zip	Phone
330	Abbie Walker	1988 Druid Hwy.	Silver City	GA	30041	6785557272

```
SQL> @qqP
```

Pid	Pname	Ptype	Qty on hand	Price	Type
2000	Paint Buckets	Paint	400	3.95	ITEM

```
SQL> @Ti
USER is "PAT2"
CURRENT_TIMESTAMP
---------------------------------------------------------------------------
22-DEC-21 08.33.58.197000 PM -06:00
```

At 08.33.58 Pat updates the **Product** table.

```
SQL> UPDATE Product set QOH = 401 WHERE product_id = 2000;
```

1 row updated.

```
SQL> @ti
USER is "PAT2"

CURRENT_TIMESTAMP
---------------------------------------------------------------------------
22-DEC-21 08.34.30.147000 PM -06:00

SQL> @ti
USER is "PAT2"

CURRENT_TIMESTAMP
---------------------------------------------------------------------------
```

22-DEC-21 08.36.41.828000 PM -06:00

Here, Pat sees the un-updated **Customer** table. Chris updated the **Customer** table at 08.35.07, but Pat does not see the update because Chris's transaction is still going on.

SQL> @qqC

Cid	Cname	Caddr	Ccity	ST	zip	Phone
330	Abbie Walker	1988 Druid Hwy.	Silver City	GA	30041	6785557272

SQL> @qqP

Pid	Pname	Ptype	Qty on hand	Price	Type
2000					

At 08.35.07 Chris updated the **Customer** table and hence locked the *Customer_id* = 330 row.

At 08.36.35 Chris attempts to update the **Product** table which is locked by Pat so Chris will wait for Pat to end transaction.

At 08.36.41 Pat issues an UPDATE on the **Customer** table, *Customer_id* = 330

Here is the deadlock:
Pat is trying to update **Customer** which was locked by Chris at 08.35.07.
Chris is trying to update **Product** at 08.36.35, but it was locked by Pat at 08.33.58.

SQL> UPDATE customer set ccity = 'Polk' WHERE customer_id = 330;
UPDATE customer set ccity = 'Polk' WHERE customer_id = 330
    *
ERROR at line 1:
ORA-00060: deadlock detected while waiting for resource

Oracle sees the deadlock and does a ROLLBACK on Pat's transaction at 08.38.47.

```
SQL> @ti
USER is "PAT2"

CURRENT_TIMESTAMP
---------------------------------------------------------------------------
22-DEC-21 08.38.47.222000 PM -06:00

SQL> @ti
USER is "PAT2"

SQL> SHOW USER
USER is "Chris2"
SQL> @ti
USER is "Chris2"

CURRENT_TIMESTAMP
---------------------------------------------------------------------------
22-DEC-21 08.33.15.313000 PM -06:00
```

At 08.33.15 Chris begins a transaction and sees the **Customer** and **Product** tables like this:

SQL> @qqC

Cid	Cname	Caddr	Ccity	ST	zip	Phone
330	Abbie Walker	1988 Druid Hwy.	Silver City	GA	30041	6785557272

SQL> @qqP

Pid	Pname	Ptype	Qty on hand	Price	Type
2000	Paint Buckets	Paint	400	3.95	ITEM

```
SQL> @ti
USER is "Chris2"
```

CURRENT_TIMESTAMP

22-DEC-21 08.33.41.162000 PM -06:00

```
SQL> SHOW USER
USER is "Chris2"
SQL> @ti
USER is "Chris2"
```

CURRENT_TIMESTAMP

22-DEC-21 08.34.51.098000 PM -06:00

We know that at 08.33.58 Pat updates the **Product** table. So, what Chris is seeing is the un-updated **Product** table because Pat has not yet ended Pat's transaction.

```
SQL> @qqP
```

Pid	Pname	Ptype	Qty on hand	Price	Type
2000	Paint Buckets	Paint	400	3.95	ITEM

```
SQL> @ti
USER is "Chris2"
```

CURRENT_TIMESTAMP

```
22-DEC-21 08.35.07.010000 PM -06:00

SQL> @qqC

 Cid   Cname           Caddr             Ccity          ST   zip     Phone
 ----  --------------- ----------------- -------------- --   ------  ----------------
 330   Abbie Walker    1988 Druid Hwy.   Silver City    GA   30041   6785557272
```

At 08.35.07 Chris updates the **Customer** table:

```
SQL> @qqP

 Pid    Pname             Ptype             Qty on hand    Price     Type
 -----  ----------------- ----------------- -----------    --------- ----------------
 2000   Paint Buckets     Paint             400            3.95      ITEM

SQL> @ti
USER is "Chris2"

CURRENT_TIMESTAMP
---------------------------------------------------------------------------
22-DEC-21 08.35.57.835000 PM -06:00

SQL> UPDATE customer SET ccity = 'Cumming' WHERE customer_id = 330;

1 row updated.

SQL> @qqC

 Cid   Cname           Caddr             Ccity          ST   zip     Phone
 ----  --------------- ----------------- -------------- --   ------  ----------------
 330   Abbie Walker    1988 Druid Hwy.   Cumming        GA   30041   6785557272
```

```
SQL> @ti
USER is "Chris2"

CURRENT_TIMESTAMP
---------------------------------------------------------------------------
22-DEC-21 08.36.35.400000 PM -06:00
```

At 08.36.35 Chris attempts to update the **Product** table which is locked by Pat -- Chris waits.

```
SQL> @qqP
  Pid   Pname           Ptype             Qty on hand   Price     Type
  ----- --------------- ----------------- -----------   --------- -----------------
  2000  Paint Buckets   Paint             400           3.95      ITEM

SQL> @Ti
USER is "Chris2"

CURRENT_TIMESTAMP
---------------------------------------------------------------------------
22-DEC-21 08.38.09.766000 PM -06:00

SQL> UPDATE Product SET QOH = 405 WHERE Product_id = 2000;

1 row updated.
```

At 08.38.09 Chris's transaction updates the **Product** table which is now allowed because Oracle terminated Pat's transaction with ROLLBACK due to the deadlock.

Here is Chris's view after updates:

```
SQL> @Ti
USER is "Chris2"
```

CURRENT_TIMESTAMP

22-DEC-21 08.39.18.344000 PM -06:00

SQL> @QqC

Cid	Cname	Caddr	Ccity	ST	zip	Phone
330	Abbie Walker	1988 Druid Hwy.	Cumming	GA	30041	678 5557272

SQL> @Qqp

Pid	Pname	Ptype	Qty on hand	Price	Type
2000	Paint Buckets	Paint	405	3.95	ITEM

Chris was able to update the **Product** table and the **Customer** table.

SQL> ROLLBACK;

Rollback complete.

Chris does a ROLLBACK and restores both the **Customer** and **Product** tables.

SQL> @Qqc

Cid	Cname	Caddr	Ccity	ST	zip	Phone
330	Abbie Walker	1988 Druid Hwy.	Silver City	GA	30041	6785557272

SQL> @Qqp

Pid	Pname	Ptype	Qty on hand	Price	Type
2000	Paint Buckets	Paint	400	3.95	ITEM

The timeline for these transactions is as follows:

At 08.33.58 Pat updates the **Product** table, P*roduct_id* = 2000.

At 08.35.07 Chris updates the **Customer** table, *Customer_id* = 330.

At 08.36.35 Chris attempts to update the **Product** table where *Product_id* = 2000, but that row is locked by Pat so Chris will wait for Pat to end transaction.

At 08.36.41 Pat issues an UPDATE on the **Customer** table, *Customer_id* = 330.

Oracle sees the deadlock and does a ROLLBACK on Pat's transaction.

At 08.38.09 Chris's transaction updates the **Product** table which is now allowed because Oracle terminated Pat's transaction with ROLLBACK due to the deadlock.

Exercises for Chapter 10

Ex 10.1.

Use two different host sign-ons to connect to Oracle with two people. The better way to do this is to use two different computers. Let one person be Van and the other Pat.

Before you begin, have Pat GRANT UPDATE on **Product** to Van and have Van GRANT UPDATE on **Vendor** to Pat

Let Pat update the **Product** table, *Product_id* = 1000 changing the Price to 9.70.

Let Van update **Vendor**, *Vendor_ID* = 200 changing Vcity to Milton

Let Pat update **Vendor**, *Vendor_ID* = 200 changing the company name to Pumps R Us

Let Van UPDATE **Product**, Product_id = 1000 changing QOH to 105

This set of transactions will deadlock. As you do these commands for Pat and Van, use the Time_now.sql script from Chapter 9 and SHOW USER before you do any command. When the two transactions are complete, ROLLBACK both to end the transactions. One will not need to be rolled back because Oracle already did it when the deadlock occurred. Which one does not need a ROLLBACK? Since each transaction involved two

UPDATEs, which UPDATE completed (check before you end the transactions explicitly).

Ex 10.2.

Modify the above scenario and re-do the exercise, but have Pat and Van explicitly LOCK the tables they intend to use first in SHARE mode and then in EXCLUSIVE mode. What changes took place in the execution of the commands?

Ex 10.3.

Modify the above scenario and re-do the exercise, but this time have Pat and Van issue SET TRANSACTION commands, one for each isolation level and granularity, i.e., once for transaction level and once for session level for each of the three isolation levels. Notice what effect each SET TRANSACTION has on the sequence of events leading to deadlock with no locking or transaction sets.

As you do this exercise, set isolation levels within each person's transaction and then again as a SET at the session level. When executing SET TRANSACION or ALTER SESION SET TRANSACTION commands, you need to do so like this:

COMMIT;
SET TRANSACTION LEVEL ...
Do whatever
COMMIT;

The reason for the COMMITs is to insure you are starting a transaction when you are testing. Of course, you need to think about what you have been doing and be sure COMMIT is appropriate for your work.

As you do these transactions as (say) Pat, note what Van is allowed to do and not do. Carefully note when Van's transaction is waiting and what happens when Pat COMMITs (or does a ROLLBACK).

REFERENCES

[10.1] https://en.wikipedia.org/wiki/ACID.

[10.2] Database concepts, Chapter 13, *Data Concurrency and Consistency*, https://docs.oracle.com/cd/B19306_01/server.102/b14220/consist.htm.

[10.2] LOCK TABLE statement:
https://docs.oracle.com/javadb/10.8.3.0/ref/rrefsqlj40506.html.

[10.3] Database SQL Language Reference, Automatic Locks in DML Operations.
https://docs.oracle.com/cloud/latest/db112/SQLRF/ap_locks001.htm#SQLRF55502.

[10.4] A higher level of granularity than rows and tables is not experienced by users -- the database itself can be locked via a process called "quiesce." The database can be put into a quiescent state by two users SYS and SYSTEM. These two users are created when the database is created and have privileges even more powerful than that of the DBA. These two users can do backup and recovery operations of the entire database as well as system upgrades. If such operations are required, then the database itself is quiesced; and basically everyone is locked out until the backup or upgrade is dealt with.

[10.5] "Oracle Isolation Level Tips," Burleson Consulting, http://dba-oracle.com/t_oracle_isolation_level.htm, retrieved April 25, 2020.

[10.6] Database SQL Language Reference.
https://docs.oracle.com/cd/B28359_01/server.111/b28286/statements_10005.htm#SQLRF01705.

[10.7] "Oracle Deadlocks," Burleson Consulting, http://www.dba-oracle.com/t_oracle_deadlock.htm, retrieved June 17, 2020.

Other Websites Related To Multi-User Topics from Which the Reader May Dig Deeper Into Various Subjects

Transactions: http://www.dba-oracle.com/t_v_transaction.htm

Database SQL Language Reference
https://docs.oracle.com/cd/B28359_01/server.111/b28286/statements_10005.htm#SQLRF01705

Burleson Consulting, Isolation Levels:
http://dba-oracle.com/t_oracle_isolation_level.htm

Burleson on dirty reads:
http://www.dba-oracle.com/t_oracle_dirty_reads.htm
Burleson on deadlocks:
http://www.dba-oracle.com/t_oracle_deadlock.htm

LOCK TABLE:
https://docs.oracle.com/javadb/10.8.3.0/ref/rrefsqlj40506.html

ACID: https://en.wikipedia.org/wiki/ACID

Setting transaction levels in SQL Server:
https://docs.microsoft.com/en-us/sql/t-sql/statements/set-transaction-isolation-level-transact-sql

https://docs.microsoft.com/en-us/sql/odbc/reference/develop-app/transaction-isolation-levels

ABOUT THE AUTHORS

Dr. Richard Walsh Earp is the former Chair of and a former Associate Professor in the Department of Computer Science and is the former Dean of the College of Science and Technology at the University of West Florida in Pensacola, Florida, USA. He has taught a variety of Computer Science courses including Database Systems and Advanced Database Systems. Dr. Earp has authored and co-authored several papers and has co-authored several books with Dr. Bagui. Some of the books co-authored with Dr. Bagui include: Learning SQL: A Step-by-Step Guide using Oracle, Database Design Using ER Diagrams, Learning SQL: A Step-by-Step Guide using Access, SQL Server 2014: A Step by Step Guide to Learning SQL, A Practical Guide to Using SQL in Oracle.

Dr. Sikha Saha Bagui is Professor and Askew Fellow in the Department of Computer Science, at The University West Florida, Pensacola, Florida. Dr. Bagui is active in publishing peer reviewed journal articles in the areas of database design, data mining, BigData, pattern recognition, and statistical computing. Dr. Bagui has worked on funded as well unfunded research projects and has over 75 peer reviewed journal publications. Dr. Bagui has also co-authored several books on Oracle SQL, SQL Server, Access SQL, and Database Design with Dr. Richard Earp. Bagui also serves as Associate Editor and is on the editorial board of several journals.

INDEX

#

1NF, 3, 4, 6, 10, 23
2NF, 4, 14, 15, 17, 18, 19, 23
3NF, 4, 19, 21, 22, 23

A

accept, 16, 32, 61, 65, 69
account, x, 26, 37, 43, 46, 47, 50, 51, 52, 59, 60, 61, 63, 65, 66, 70, 95, 143
acid, viii, 145, 146, 147, 167, 168
add constraint, 29, 31
after update, 105, 106, 107, 108, 109, 163
alter, 27, 29, 30, 31, 37, 39, 46, 67, 84, 118, 125, 152, 155, 166
alter table, 27, 29, 30, 31, 37, 39
alter trigger, 118
anomaly, 17, 18, 150
atomic, 4, 7, 9, 10
attribute, x, 7, 8, 9, 10, 11, 13, 15, 17, 19, 27, 28, 29, 30, 32, 33, 34, 40, 76, 100, 102, 110, 116, 117, 119
AVG, 82

B

before update, 105

C

candidate key, 11, 13, 14
cascade, 36, 39, 60, 61, 65, 69, 99
check constraint, 29, 32, 38, 100
commit, viii, 124, 125, 126, 127, 130, 133, 134, 142, 143, 147, 148, 149, 166
composite attribute, 4, 9
composite data, 8
concatenated primary key, 28
concatenation, 11, 12, 19
concurrency, viii, 145, 146, 147, 148, 151, 153, 156, 167
conn, 49, 65, 66, 67, 93, 128
connect, 44, 46, 47, 48, 49, 52, 62, 65, 66, 70, 84, 85, 93, 95, 101, 102, 127, 141, 142, 165
constraints, viii, ix, 26, 29, 30, 31, 32, 33, 37, 38, 40, 63, 84, 89, 91, 93, 99, 100

count, 23, 70, 73, 74, 75, 82, 98, 99, 135, 136, 137, 138, 140
create, x, 26, 27, 28, 30, 32, 33, 36, 37, 38, 39, 45, 46, 47, 48, 49, 50, 51, 52, 53, 54, 55, 56, 57, 58, 59, 60, 61, 62, 63, 65, 66, 67, 68, 69, 70, 78, 79, 81, 82, 84, 89, 90, 91, 93, 94, 95, 100, 101, 102, 107, 108, 109, 111, 112, 113, 116, 118, 119, 125, 141, 142
create or replace trigger, 107, 108, 109, 112
create session, 45, 47, 48, 49, 52, 53, 60, 62, 69
create synonym, 66, 67, 68, 69, 81
create table, 26, 27, 28, 30, 32, 33, 36, 37, 39, 46, 48, 49, 50, 52, 53, 59, 60, 62, 63, 69, 79, 81, 90, 91, 93, 95, 100, 107, 108, 111, 113, 116
create tablespace, 50
create trigger, 107
create user, 46, 47, 48, 49, 50, 51, 52, 53, 54, 59, 60, 61, 62, 65, 69

D

data dictionary, 52, 73, 77, 79, 127
database, vii, viii, ix, x, 1, 2, 3, 4, 5, 6, 10, 11, 12, 13, 14, 16, 18, 19, 22, 23, 25, 26, 27, 29, 34, 40, 41, 42, 43, 44, 45, 46, 47, 48, 50, 51, 52, 59, 62, 65, 66, 70, 81, 84, 87, 89, 93, 94, 97, 98, 99, 102, 105, 118, 119, 120, 121, 122, 124, 125, 130, 131,132, 133, 142, 145, 146, 147, 148, 149, 152, 153, 156, 167, 168, 169
database administrator, 42, 43, 44, 45
database integrity, vii, 1, 25, 26
database objects, 45
DBA, 42, 43, 44, 45, 47, 50, 51, 52, 56, 66, 69, 77, 79, 83, 89, 127, 153, 167
DDL, 125, 155
deadlock, 156, 157, 159, 160, 163, 165, 166, 167, 168

decomposition, 6, 21
default tablespace, 50, 51, 52, 54, 61, 65
delete, 18, 34, 35, 36, 39, 46, 83, 84, 85, 86, 93, 98, 99, 102, 105, 106, 111, 112, 114, 115, 119, 121, 130, 131, 155
desc, 57, 73, 74, 75, 88, 135, 139
developers, 42, 45, 46
dictionary, vii, 29, 57, 73, 74, 76, 77, 78, 79, 85, 127
dirty reads, 149, 151, 152, 168
disable, 29, 30, 31, 37, 118, 156
disable constraint, 31, 37
DML, viii, 81, 83, 85, 100, 118, 121, 125, 127, 130, 132, 133, 146, 147, 153, 154, 167
drop, 29, 30, 31, 37, 46, 60, 61, 63, 65, 69, 79, 125, 155
drop constraint, 31, 37
drop table, 63, 79, 155
drop user, 60, 61, 65
durability, 146, 147

E

enable, 29, 30, 31, 36, 37, 118
enable constraint, 31, 37
equi-join, 22
execute, 39, 47, 48, 53, 57, 58, 65, 66, 69, 70, 81, 84, 85, 100, 101, 119, 130, 141, 142, 147
exit, 53, 54, 57, 94

F

files, 2, 44
first normal form, 3, 4, 6
foreign key, 33, 34, 35, 36, 38, 89, 100, 103
functional dependencies, 12, 20, 21, 22

G

grant, viii, 46, 47, 48, 49, 50, 51, 52, 61, 62, 65, 66, 67, 68, 69, 79, 83, 84, 93, 95, 107, 127, 165
grant all, viii, 83, 84
grant create session, 47, 49, 61, 65
grant create synonym, 66, 67
grant select, 65, 68, 83, 95, 127
grant update, 165
group by, 82, 135, 136, 138, 140

H

having, 4, 11, 33, 47
help set, 55, 57, 58
hierarchy of users., 46, 50
host, 50, 53, 54, 56, 57, 63, 64, 94, 165

I

identified by, 19, 20, 27, 46, 47, 49, 51, 52, 54, 61, 65
index, viii, 44, 57, 74, 171
insert, 18, 34, 35, 36, 39, 45, 63, 64, 79, 83, 84, 85, 86, 91, 92, 93, 95, 98, 101, 102, 105, 106, 107, 108, 109, 111, 112, 114, 115, 116, 117, 119, 121, 135, 137, 139, 150, 155
insert into, 45, 63, 64, 79, 86, 91, 92, 95, 107, 108, 109, 112, 114, 116, 117, 119, 135, 137, 139
isolation, 146, 148, 149, 150, 151, 152, 153, 154, 166, 167, 168
isolation level, 148, 149, 150, 151, 152, 153, 166, 167, 168

K

key, 10, 11, 12, 13, 14, 15, 17, 18, 19, 21, 27, 28, 29, 34, 35, 36, 57, 87, 89, 103

L

linesize, 55, 56, 57, 58, 114, 122
lock, 103, 147, 151, 153, 154, 155, 156, 157, 166, 167, 168
lock table, 153, 154, 155, 167, 168

M

M:N relationship, 19, 87, 89, 91
mathematical sets, 7, 11
max, 75, 82, 97, 102
min, 75, 82, 101, 102
minimal key, 12, 13
multi-user environment, ix, 26, 43, 44, 46, 146, 147

N

normal form, ix, 2, 3, 4, 6, 12, 15, 19, 22, 24
normalizing, 4
not null, 29, 30, 32, 33, 35, 36, 37, 40, 63, 75, 100
null, 18, 19, 29, 30, 33, 35, 36, 39, 74, 75, 88, 117, 135, 139

O

object, 43, 45, 46, 60, 83, 84, 156
oracle, i, iii, ix, x, 29, 31, 40, 46, 52, 57, 58, 93, 103, 117, 119, 120, 124, 125, 130, 142, 143, 147, 148, 149, 150, 151, 152, 153, 154, 155, 156, 157, 160, 163, 165, 167, 168, 169
order by, 47

P

password, 46, 51, 52, 54, 56, 61
phantom reads, 149, 152
primary key, 11, 12, 13, 14, 15, 18, 19, 20, 22, 27, 28, 29, 30, 31, 32, 33, 34, 35, 36, 38, 39, 40, 63, 89, 100
privilege, 43, 45, 46, 47, 48, 49, 50, 52, 53, 61, 62, 66, 67, 68, 78, 84, 85, 102, 107, 154, 157
prompt, 55, 56, 60, 61, 65, 69, 99
public, 68

Q

quota, 51, 52, 54, 61, 65

R

read committed, 148, 149, 150, 151, 152
read only, 148, 152
read uncommitted, 148, 149, 151
redundancy, 17, 19
references, vii, viii, 3, 24, 34, 36, 40, 58, 84, 100, 103, 119, 143, 167
referential integrity, 32, 33, 34, 35, 36, 99
relational database, i, iii, ix, 2, 3, 4, 6, 7, 11, 13, 21, 33
repeating attribute, 22
repeating groups, 4, 5, 6, 10
restrict, 36, 99
revoke, 50, 68, 85
role, 51, 60, 61, 66, 78, 79
rollback, viii, 124, 125, 127, 129, 130, 131, 133, 134, 137, 141, 142, 143, 146, 147, 148, 149, 155, 160, 163, 164, 165, 166
row, 4, 5, 6, 7, 8, 9, 11, 12, 18, 19, 27, 33, 34, 35, 58, 85, 105, 106, 107, 108, 110, 111, 112, 114, 117, 119, 123, 127, 128, 130, 135, 137, 139, 140, 146, 148, 150, 153, 154, 155, 156, 157, 158, 159, 162, 163, 165
row-level, 105, 106

S

script, 53, 54, 55, 56, 57, 58, 60, 61, 62, 63, 64, 65, 69, 78, 79, 82, 84, 85, 86, 91, 94, 95, 101, 102, 109, 112, 114, 122, 123, 124, 128, 129, 142, 156, 165
security, ix, 26, 27, 28, 30, 41, 44, 45, 68
select, 22, 39, 47, 52, 57, 58, 64, 65, 68, 70, 73, 74, 75, 76, 79, 81, 82, 84, 85, 86, 89, 90, 92, 93, 94, 95, 96, 97, 100, 101, 110, 113, 114, 116, 122, 123, 124, 126, 127, 128, 129, 134, 135, 136, 137, 138, 139, 140, 141, 155, 157
serializable, 121, 145, 146, 148, 152
serialized transaction, viii, 121, 145, 146
set, 2, 3, 7, 8, 11, 12, 26, 36, 38, 39, 44, 45, 53, 55, 56, 57, 58, 61, 64, 73, 74, 77, 79, 87, 89, 99, 106, 110, 114, 118, 122, 123, 127, 128, 137, 140, 147, 150, 151, 152, 158, 159, 162, 163, 165, 166, 168
set null, 36, 39, 99
set rownum, 57, 58
set transaction, 151, 152, 166
set wrap off, 56, 57, 122
show user, 57, 122, 123, 124, 126, 127, 129, 141, 157, 160, 161, 165
SQL, ix, 26, 28, 40, 41, 42, 47, 49, 50, 52, 53, 54, 55, 56, 57, 58, 63, 64, 65, 66, 67, 68, 71, 74, 75, 76, 85, 88, 90, 92, 93, 94, 95, 96, 105, 109, 110, 113, 114, 115, 116, 118, 119, 121, 122, 123, 124, 126, 127, 128, 129, 131, 134, 135, 136, 137, 138, 139, 140, 141, 157, 158, 159, 160, 161, 162, 163, 164, 167, 168, 169
SQL*plus, 53, 55, 56
starvation, viii, 156

Index

statement, 29, 36, 46, 48, 64, 77, 84, 105, 106, 118, 119, 121, 125, 130, 132, 151, 153, 156, 167
statement-level, 105, 106
SUM, 82, 102
synonyms, 45, 70, 89
system administrator, 41, 42, 50, 51, 56, 69
system privileges, 43, 45, 46

T

table, x, 3, 4, 5, 6, 7, 8, 9, 10, 11, 12, 13, 14, 15, 16, 17, 18, 19, 20, 21, 22, 26, 27, 28, 29, 30, 31, 32, 33, 34, 35, 36, 37, 38, 39, 40, 43, 45, 46, 49, 52, 55, 57, 59, 62, 63, 64, 65, 66, 68, 69, 73, 74, 75, 76, 77, 78, 79, 81, 83, 84, 85, 86, 87, 89, 90, 91, 93, 94, 95, 96, 98, 99, 100, 101, 102, 103, 105, 106, 107, 108, 109, 110, 111, 112, 113, 114, 115, 116, 118, 119, 121, 124, 126, 127, 130, 131, 132, 133, 134, 137, 141, 142, 146, 147, 148, 149, 150, 151, 153, 154, 155, 156, 157, 158, 159, 161, 162,163, 164, 165
table_privileges, 78, 79
tablespace, x, 43, 50, 51, 54, 56, 60, 75, 76
temporary user, 46, 60
third, ix, 2, 4, 5, 19, 24, 29
to_date, 90, 91, 92, 135, 137, 139
transaction, 90, 91, 93, 102, 121, 124, 125, 126, 127, 128, 129, 130, 131, 132, 133, 134, 135, 141, 142, 143, 145, 146, 147, 148, 149, 150, 151, 152, 153, 154, 155, 156, 157, 159, 160, 161, 163, 165, 166, 168
transitive dependency, 21

trigger, viii, 105, 106, 107, 109, 111, 112, 113, 116, 117, 118, 119

U

unique constraint, 33
unique key, 100
unix, 53, 55, 142
update, 6, 18, 35, 36, 39, 45, 83, 84, 85, 93, 98, 99, 101, 102, 105, 106, 107, 108, 109, 110, 111, 112, 114, 115, 119, 121, 123, 124, 127, 128, 129, 132, 137, 140, 142, 146, 148, 152, 154, 155, 156, 157, 158, 159, 162, 163, 164, 165, 166
upper, 75

user_tables, 75, 76, 79
user_tablespaces, 75, 76
user-defined, 51, 79
username, 46, 51, 52, 53, 56, 128, 141
users, viii, ix, x, 25, 26, 41, 42, 43, 44, 45, 46, 47, 48, 50, 52, 53, 55, 59, 60, 61, 62, 63, 65, 69, 77, 78, 81, 83, 84, 86, 87, 89, 93, 94, 98, 101, 105, 107, 118, 121, 145, 147, 151, 153, 154, 156, 167

W

where, x, 2, 7, 8, 9, 11, 12, 14, 15, 21, 22, 27, 30, 34, 43, 47, 51, 59, 75, 77, 79, 83, 96, 97, 99, 103, 110, 114, 117, 123, 124, 126, 127, 128, 129, 130, 131, 134, 135, 136, 137, 138, 140, 141, 146, 156, 157, 158, 159, 162, 163, 165
with admin option, 48, 49, 52, 61, 62, 66, 67
wrap, 55, 56, 57, 58
wrap on, 55, 57